Care Giving for Alzheimer's Disease

Verna Benner Carson • Katherine Johnson Vanderhorst • Harold G. Koenig

Care Giving for Alzheimer's Disease

A Compassionate Guide for Clinicians and Loved Ones

Springer

Verna Benner Carson
C&V Care Specialists, Inc.
Fallston
Maryland
USA

Katherine Johnson Vanderhorst
C&V Care Specialists, Inc.
Williamsville
New York
USA

Harold G. Koenig
Dept. of Psychiatry Behavioral Science
Duke University Medical Center
Durham
North Carolina
USA

ISBN 978-1-4939-4527-6 ISBN 978-1-4939-2407-3 (eBook)
DOI 10.1007/978-1-4939-2407-3

Springer New York Heidelberg Dordrecht London

Softcover reprint of the hardcover 1st edition 2015

Printed on acid-free paper

Springer New York is part of Springer Science+Business Media (www.springer.com)

Foreword

In this excellent and informative book Verna Carson introduces the public to the concepts of retrogenesis and the importance of these concepts for understanding persons with dementia, especially the dementia of Alzheimer's disease. Retrogenesis has been defined as "the process by which degenerative mechanisms reverse the order of acquisition in normal human development." As Verna points out in this book, this process has long been recognized in a certain sense, for example, by some of the most famous playwrights of all time. Even before the playwrights, a famous mathematician and philosopher who gave the world some of the "a, b, c's" of mathematics, also provided the "a, b, c's" of life. In the sixth century B.C., Pythagoras famously stated that for a right angle triangle, the square of the abscissa (a^2), plus the square of the ordinate (b^2), equals the square of the hypotenuse (c^2) or "$a^2+b^2=c^2$." All of us who studied geometry in high school have learned these truths, still called "the Pythagoras theorem." However, what Pythagoras may ultimately be known for is having provided us with the "a, b, c's of life." As a philosopher, Pythagoras stated that there are five phases of life. He said that the last two phases of life are "the senium" in which the brain returns to the first epoch of its infancy.

In the 1980s, I, together with my associates, systematically described the stages of brain aging and Alzheimer's disease. These descriptions have been proven to be scientifically valid, and also very useful. For example, our Global Deterioration Scale is used by the Alzheimer's Association to help the public understand the course of brain aging and Alzheimer's disease (www.alz.org/AboutAD/Stages.asp). Shortly after we developed the Global Deterioration Scale, we realized that functionally it was possible to describe the course of progressive brain aging and Alzheimer's disease in particular detail and we showed that a total of 16 successive "functional" stages of brain aging and Alzheimer's disease could be described with a measure called the "FAST," the Functional Assessment Staging Tool procedure. In the process of developing the FAST, we recognized that these functional stages are a very precise reversal of the order of acquisition of the same stages in normal human development (Reisberg, *Geriatrics*, 1986).

With these insights we set out to answer three questions: (1) In what other ways does the aging and Alzheimer's disease process reverse normal human development?

(2) How can these insights help in the care of persons with Alzheimer's disease? and (3) How do these discoveries relate to the cause of Alzheimer's disease?

The first and foremost question that we addressed was: do thinking abilities also reverse normal development, and if so, how close are the parallels? Working with investigators in Japan, especially Kenichi Meguro and Masumi Shimada, we showed close relationships between decline in Alzheimer's disease on a well-known intelligence test and the progression of functional loss on the FAST, and that these losses closely mirrored the corresponding developmental ages (Shimada and colleagues, *Psychogeriatrics*, 2003). Also, because in the later portion of the FAST scale persons with severe AD are untestable on traditional tests designed for children, we took infant intelligence test measures and adapted them for severe Alzheimer's disease patients. When we tested the severe Alzheimer's disease patients we found a very close correspondence between the performance in the severe Alzheimer's disease patients and that of infants at the corresponding points of Alzheimer's functional losses and infant functional attainment (Sclan and colleagues, *Psychiatric Journal of the University of Ottawa*, 1990 and Auer and colleagues, *Journal of the American Geriatric Society*, 1994). Most recently, investigators in Spain led by Jordi Peña-Casanova and Rubén Muñiz took the final, conclusive step. They, directly compared Spanish school children from ages 4 to 12 with elderly persons with normal aging, mild cognitive impairment, or progressively more severe Alzheimer's disease. They found strikingly strong parallels between the losses in aging and Alzheimer's disease and the corresponding acquisition patterns in children for both thinking (cognition) measures and functioning measures (Rubial-Álvarez and colleagues, *Journal of Alzheimer's Disease*, 2013).

Also, we and others showed that in many other interesting ways, the progression of Alzheimer's disease reverses the normal human development process. For example, we showed that reflexes in infants emerge in Alzheimer's patients at the corresponding developmental age related Alzheimer's disease stages in a series of publications by Emile Franssen and our other colleagues. Also, in terms of brain structure (anatomy), the process of brain aging has been shown to reverse the normal human development pattern. For example, the brain contains cells called "neurons" which have extensions similar to arms or wires called "axons" that are used to communicate with other brain cells. These axons (the nerve cell wires) are progressively covered with a fat-like substance, similar to the rubber which insulates a wire, and which aides nerve cell communication with other nerve cells. The fat-like substance on the nerve extensions is called "myelin." The process of nerve cell "rubberization" called "myelination," continues from infancy to late life. Several early investigators in Alzheimer's disease brain changes noted that the unmyelinated and most recently and therefore, most thinly myelinated brain regions are the first to be affected by Alzheimer's type brain changes. These early investigators included Arne Brun, Elisabet Englund and Lars Gustafson in Sweden, Patrick McGeer in Canada, and Heiko Braak in Germany. With the advent of new brain imaging techniques called diffusion tensor imaging, we were able to begin to directly verify the validity of this process in Alzheimer's persons (Choi and colleagues including Kelvin Lim and Isabel Monteiro, *Journal of Geriatric Psychiatry and Neurology*, 2005).

Subsequently, a large study in several US centers strongly supported these observations (Stricker and colleagues, *Neurology*, 2009). There is also evidence that other brain changes, especially the loss of brain energy, called metabolism, also reverses the normal brain developmental patterns. Building on these observations we (especially Sunnie Kenowsky, Stefanie Auer and I), developed a science of Alzheimer's disease management and treatment based upon the knowledge of the developmental age (DA) of the Alzheimer's disease (AD) person (Reisberg and colleagues, *International Psychogeriatrics*, 1999 and Reisberg and colleagues, *American Journal of Alzheimer's Disease and Other Dementias*, 2002). With this management science, we translated the AD stages into correspondingly developmental ages. We have shown that the management needs and care requirements of AD persons at the AD stages mirror the corresponding DAs. Many of the emotional changes, activity needs, and other aspects of AD persons also reflect the DA. We have also noted differences between AD persons and their DA "peers." For example AD persons do not undergo a physical retrogenesis, so they are, for example, larger and potentially stronger than children at the same DA. Also, AD persons are older and prone to the illnesses (comorbidities) of older persons, unlike children. Additionally, Alzheimer's persons have a history, for example, of speaking, which infants do not. Hence, the "science of AD management" includes differences in the model as well as similarities. Recently, in part by applying this retrogenesis model to AD person's management and care, Dr. Kenowsky and I, have been getting very dramatically positive results in persons with moderately severe AD (Reisberg, et al., *Neuropsychopharmacology*, 2013 [abstract]).

Finally, I believe these findings are moving us towards a better understanding of the "cause" of AD. Basic brain mechanisms including brain anatomy (structure) and physiology (energy utilization), appear to be reversing development. In many ways we have found that even the time course of AD losses reverses normal developmental attainments, although there are also some differences. These observations point to a brain developmental reversal, perhaps, fueled by losses in brain energy (i.e., metabolic decrements). Accordingly, conditions which interfere with brain energy utilization (for example, condition which are diabetogenic), such as obesity, inactivity, etc., are increasingly being found to be risk factors for AD. Ultimately, the cause of AD, of course, requires further investigation. In the interim, advances in our understanding of AD persons can help these increasingly needy and potentially increasingly disturbed AD persons, as the disease process advances, today. Because of these possibilities and findings, this year, in 2014, the National Alzheimer Society of Spain (CEAFA) implemented a care model requiring familiarity with the use of our FAST scale of the progression of Alzheimer's disease related functional losses, and the retrogenic concepts of care needs and care possibilities, as compulsory in all nursing homes in Spain.

Director, Fisher Alzheimer's Disease Program
Clinical Director, Aging & Dementia
Clinical Research Center
New York University Langone Medical Center

Barry Reisberg, M.D.
Professor of Psychiatry

Preface

This book is written by three providers who care deeply about and are intimately involved with caregivers who struggle on a daily basis to provide loving and patient care to those diagnosed with Alzheimer's disease or one of the other dementias. Dr. Harold G. Koenig is a professor, a geriatric psychiatrist, and a researcher who delivers care to patients afflicted with one of the many dementias, while at the same time providing support to the caregivers of these individuals. Dr. Verna Benner Carson is a psychiatric nurse and President of a consulting company, C&V Senior Care Specialists, Inc. She developed a program entitled *Becoming an Alzheimer's Whisperer* and along with her business partner, Katherine Johnson Vanderhorst, also a psychiatric nurse, they traverse the country training family as well as professional caregivers—nurses, occupational therapists, physical therapists, and social workers—to "think outside of the box" when trying to manage the challenging behaviors of those with dementia.

They accomplish this by teaching and applying the Theory of Retrogenesis, developed by Dr. Barry Reisberg. This theory posits that those afflicted with dementia regress in a backwards fashion from adulthood to infancy. The theory does not suggest that a caregiver should "talk down" to the person with Alzheimer's or use "baby talk" but it does guide caregivers towards an accurate appraisal of what the individual is and is not capable of doing. Anyone who has seen or cared for an individual in the end stage of Alzheimer's cannot ignore the fact that the individual has lost many of the skills we associate with being an adult—the person is in a fetal position, cannot speak, lift his/her head up, sit up or walk, is incontinent and requires complete care—very much like an infant. The theory allows caregivers to understand the challenging behaviors in a different and more acceptable manner and leads to a problem solving approach that opens up a wide range of interventions. Additionally, the theory is correlated with damage to specific areas of the brain, so that caregivers learn that when their loved one repeats the same question over and over again they do not intend to be troublesome but this repetition occurs because the hippocampus, the storehouse of short term memory, is broken. Once caregivers know and understand the physiological origin of a behavior, they are free to "think outside of the box" to arrive at behavioral strategies.

Our hope is that within this book there are enough examples of such creative thinking that caregivers feel encouraged to allow their own inventiveness to lead them to discover interventions that are loving, patient, and effective in managing the challenging behaviors that characterize Alzheimer's and other dementias.

Contents

Chapter 1
Going Back to the Beginning: The Theory of Retrogenesis

As baby boomers age, the projected number of those who will develop Alzheimer's is staggering—not only for family caregivers who might still be raising children but also for the health-care system of the USA. Consider these statistics from the 2014 edition of *Alzheimer's disease Facts and Figures, Alzheimer's and Dementia* (Alzheimer's Association 2014; Auer et al. 1994; Bobinski et al. 1996).

One in nine (11 %) of those 65 and older has the disease.
One third of those 85 and older (32 %) have the disease.
Eighty two percent of those with Alzheimer's disease are 75 or older.

It is predicted that between 2010 and 2030, the number of older Americans will double to a staggering 72.1 million people, and by 2050 the number of older Americans will reach 88.5 million! For the first time in our history, people over 65 will outnumber children under five (Penny Wise, Pound Foolish: Fairness and Funding at the National Institute on Aging 2011).

Every level of health-care provider, primary care as well as specialty physicians, nurses, nurse practitioners, social workers, occupational therapists, physical therapists, and speech and language pathologists will be challenged to meet the needs of persons diagnosed with Alzheimer's. Additionally, these providers will be called on to meet the needs of family caregivers who are desperate for strategies to manage challenging behaviors and to keep a promise that is so often made to the elderly family member, "I will never place you." How will we meet these needs? What advice will we give to that struggling daughter, wife, or husband who says, "I don't know if I can do this anymore! My wife pushed me down in the bathroom. I needed stitches to the gash that I had on my scalp. I know I promised that I would never place her but I never counted on this behavior and I don't know how to manage it. Help me!"

You are about to be introduced to an innovative approach based on the *theory of retrogenesis* developed by Dr. Barry Reisberg (Reisberg 1988; Reisberg and Franssen 1999; Reisberg et al. 2002). This approach called *Becoming an Alzheimer's Whisperer* is "gentle and loving" easy to learn, and based on concrete evidence that will absolutely help you answer the desperate pleas for help that you hear from family caregivers.

V. Benner Carson et al., *Care Giving for Alzheimer's Disease*,
DOI 10.1007/978-1-4939-2407-3_1

The theory of retrogenesis is a stage theory that describes changes in the person's cognitive and functional abilities as he/she progresses through Alzheimer's disease. The theory posits that the brain of the person afflicted with Alzheimer's deteriorates in the reverse order that the brain developeds from birth. That is, the last area to be fully developed, or myelinated, is the first area to be damaged in Alzheimer's. The phrase "returning to a second childhood" and the word "dotage" defined in part as "childishness of old age" are captured in this theory. These changes can be understood in light of where the person is in her/his decline. Simply put, a person at stage 6, for instance, is functioning at the level of a toddler. Caregivers should never use the knowledge of the functional assessment staging tool (FAST) scale to justify talking to someone with Alzheimer's in a belittling manner but should adjust expectations and activities to the person's cognitive and functional level. In Chap. 2 you will read how this damage correlates not only to the function of specific areas of the brain but also to specific challenging behaviors.

The idea that some people "go backwards" in Alzheimer's is not a novel idea. As early as 423 BC, Aristophanes wrote a play in which he noted that "old men are like children twice over" (Aristophanes 1938). In Shakespeare's play, *All the World's a Stage"* Jacques's soliloquy compares the extreme old age of a man to a "second childhood." He loses control over his senses and becomes dependent on others just like a child (Agnes Latham 1967). In the early 1960s the term "returning to a second childhood" was a way of describing Alzheimer's. During that time, people who displayed the symptoms of what we now know as Alzheimer's disease were said to have "senile dementia" or "organic brain syndrome" and might be placed in one of the large state psychiatric hospitals until they died. But even much earlier than the 1960s, the disease of Alzheimer's (although not called Alzheimer's) was known. The number one risk factor is living into old age—and as health care improves we see that on average people are living much longer than they did in the mid-twentieth century.

Where did the name "Alzheimer's come from? It was the name of a German physician, Alois Alzheimer, who pioneered in linking specific behavioral symptoms to microscopic brain changes. He treated Auguste D., a patient who suffered from profound memory loss; she harbored unfounded suspicions about her family and a host of other worsening psychological changes. When she died, Dr. Alzheimer performed an autopsy. He noted dramatic shrinkage of her brain and abnormal deposits in and around nerve cells. He was a friend of Dr. Emil Kraepelin, a German psychiatrist, who worked with Dr. Alzheimer. When Dr. Kraepelin wrote the eighth edition of his book *Psychiatrie,* he named the disease documented by his friend as Alzheimer's disease.

In the late 1980s, Dr. Barry Reisberg began his work on the *theory of retrogenesis*. He published the FAST in 1988; this is an assessment tool that allows for specific identification of where someone is in her/his progression through Alzheimer's. Not only does the tool allow us to identify the appropriate stage of progression but also to know the appropriate interventions for each stage.

Let us take a look at the stages of the FAST scale and how we should respond to each stage (Reisberg 1988; Franssen et al. 1993; Reisberg and Franssen 1999).

Table 1.1 Dementia Care Specialists. (http://www.crisisprevention.com/Resources/Article-Library/Dementia-Care-Specialists-Articles/The-Adapted-FAST-Introduction-and-Application. Accessed, June 1st, 2014)

Stage on FAST	Scale Level of Cognition/ Dementia	Abilities at Each Stage	Response of Caregiver
Stage 1	Normal adult	Abilities consistent with expectations for a normal adult	Response is to a normal adult
Stage 2	Normal older adult	Very mild memory loss	Response is to a normal adult
Stage 3	Early dementia or mild cognitive impairment (MCI)	Early dementia—mild cognitive impairment	Memory problems apparent to coworkers and family members; individual with memory problems denies that he/she has a problem Stage lasts approximately 7 years
Stage 4	Mild dementia	Mild dementia functional level 8–12 years of age "great foolers" still looks good; can engage in a conversation; careful listening may reveal the holes in this person's memory	Asks the same question repeatedly, has difficulty managing finances—easily taken advantage of financially, driving skills deteriorating—might get lost even in familiar places—seek doctor's advice regarding driving—obtain a driving evaluation from occupational therapist, based on OT's findings doctor may need to send a letter to Motor Vehicle Administration based on result of OT's driving evaluation; Experiences difficulty in shopping for groceries and in preparing meals for guests. Has difficulty writing the correct date and amount on checks. May also find it difficult to order from a menu in a restaurant and might defer to spouse to order for both. Person with AD is usually in profound denial Stage lasts approximately 2 years
Stage 5	Moderate dementia	Functional level 5–7 years of age Should not live alone	Needs assistance with dressing—may need help in sequencing clothing, needs oversight for appropriateness of dress for weather and/or occasion; can participate in personal care; responds to music; likes to do repetitive behaviors, i.e., clipping coupons, folding laundry, sorting coins, likes to sing and listen to music of his/her era Can no longer live independently—someone needs to provide adequate and appropriate food, pay the rent and utilities and handle finances Can inconsistently recall information such as name of current president, current address Stage lasts approximately 1.5 years

Stage 6	Moderately severe dementia	Functional Level 4 years to—24/36 months	About 5 min of short-term memory; can still read—a powerful remaining ability; Needs help with all activities of daily living, loses ability to dress self independently and perform any personal hygiene tasks—lose the ability to brush teeth and handle mechanics of toileting; will become incontinent- first of urine then of stool, resists bathing—need to find bathing approaches other than
			showering, still enjoys engaging in repetitive behaviors, challenging behaviors become more apparent—maybe due to untreated pain—music very powerful! Challenging behaviors become more prevalent—mean duration of this stage 2–5 years
Stage 7	Severe dementia	End stage-functional level less than 24 months—deteriorating to 4–12 weeks old Stage 7C—eligible for hospice care under Medicare	7a.Only has six or fewer words 7b.Speech limited to the use of a single intelligible word—may be repeated over and over again 7c. Hospice eligible under Medicare—cannot walk without personal assistance 7d. Ability to sit up without assistance is lost-will fall over if there are not lateral rests (arms) on the chair 7e. Cannot smile

AD Alzheimer's disease

Let us take a look at some general caregiving strategies for each stage of the FAST.

During stage 3, many people who are experiencing mild cognitive impairment are too frightened to have their memory evaluated. Even when family and friends comment on the individual's failing memory, profound denial kicks in and the person can provide explanations for not remembering, i.e., "I was overly tired and that's why I forgot."; "I was preoccupied and didn't hear the directions."; "You mumbled and that is why I didn't do what you asked me to do." Any explanation for a faulty memory—other than the explanation of Alzheimer's will do. This denial is so powerful that most people at this stage avoid seeking evaluation from a physician.

During stage 4, the person still retains good social skills and can convince anyone who does not look too closely that he/she is okay. She or he might be called "a great Fooler"! The story of Mildred illustrates her ability to "fool" even the physician.

> Mildred went to see her primary care doctor for a routine checkup and to have her insulin level evaluated. Mildred was accompanied by her daughter who had taken her mother to the lab to get her lab work completed. The results of a recent HbA1c test were not encouraging to the doctor. He said, "Mildred, what is going on? Your blood work is not what it should be. Are you eating more sugar or sweets than you should be eating?" He said this with a smile but was quite alarmed with the lab results. The doctor expressed concern that Mildred's test results showed a high level of blood sugar indicating that Mildred's diabetes was not being adequately controlled. He asked Mildred to tell him what she had eaten for breakfast. Without hesitation, Mildred spouted off a perfect diabetic meal. She told him that

she had four ounces of orange juice, a half of a cup of Special K© cereal, about a half a cup of skimmed milk and a piece of toast with just a little butter. The doctor was perplexed and said he would like to send a home healthcare nurse into see Mildred and provide him with more information. The home care nurse called Mildred the evening before the scheduled visit and told Mildred that the nurse would come to her home at 7AM and not to eat or take her insulin until the nurse arrived. The nurse told Mildred that she wanted to check Mildred's glucometer and to get a good reading of her blood sugar. When the nurse arrived the next morning, Mildred couldn't produce her glucometer and told the nurse that she had already tested her blood sugar and it was fine but she was unable to elaborate on what "fine" meant. The nurse than said let's take a look at your refrigerator and your pantry to see if there are foods that you are eating that are raising your blood sugar levels. The nurse found no Special K, no orange juice or skim milk—instead she found a bag of chocolate candies and sugared drinks in Mildred's refrigerator and a gallon of ice cream in the freezer with a spoon stuck in the ice cream! Was Mildred lying to her doctor? No she was telling him what she used to eat when she actively managed her diabetes—she doesn't remember what she ate for breakfast. Mildred is a "great Fooler"! She looks good; she can engage in conversation; but what she reports might be very different from reality!

During stage 4, the ability to manage financial matters is impaired and individuals at this stage of the disease are easy targets for unsavory people who are willing and able to take advantage of the trusting nature of an elderly person whose cognitive ability is that of a child between 8 and 12 years old. Scams include asking for money for needy children but going beyond just a request for money—telling the old person that if they want to make a continuing contribution to whatever cause is being extolled, all the older person needs to do is to provide her/his bank account and routing number and automatic deductions can be made from the account. The unsavory person now has access to this trusting person's checking account. The following are signs that the person with Alzheimer's has been a victim of a financial scam:

- The person seems afraid or worried when he or she talks about money.
- Sums of unaccounted for money are missing from the person's bank or retirement accounts.
- Signatures on checks or other papers do not look like the person's signature.
- The person's will has been changed without his or her permission.
- The person's home is sold, and he or she did not agree to sell it.
- Things such as clothes or jewelry are missing from the home.
- The person has signed legal papers without knowing what the papers mean (NIH Senior Health Gov 2012).

It is so important for health-care providers to educate families to the financial risks associated with Alzheimer's disease. Some concrete strategies for health-care providers to communicate include the following:

1. Put your loved one on the no solicitation list.
2. Encourage widowed women to maintain the telephone listing in the husband's name or initial.
3. Talk to the parent of concern and attempt to get agreement that no financial decisions should occur without a discussion with someone in the family.
4. Obtaining a power of attorney (POA) is a must but in reality, POA does not prevent seniors from handling , giving away, or doing whatever he/she chooses

to do with his/her own money. Only *conservatorship*, obtained through the local Probate Court (easily done if doctor or psychologist recommend it), takes about 10 min to take care of in a private court. The person seeking financial control completes forms provided by the court and for a small fee the family member can gain absolute control of finances and is able to legally prosecute or go after any fraudulent financial scams. Gaining *conservatorship* does not preclude the older person from being involved in his/her own finances as much as is possible, but it provides a safety net that protects the vulnerable person from financial exploitation.

5. Give parent or other loved one with Alzheimer's a prepaid credit card or set low limits on current ones. And keep two checking accounts—one for senior to use and one that the responsible family member uses to pay the bills.
6. The older person needs to have some "free spending" so that he/she is not made to feel like a child.
7. The family needs to confront the reasons that their loved one is such an easy target. Many individuals with Alzheimer's are very lonely. Anyone, even a scam artist, who makes this elder feel good and/or that he/she is contributing to another in need, will have the ability to persuade and manipulate that elderly person—especially when he/she has Alzheimer's disease. Families need to make frequent visits, to make sure that the elder person gets out of the house, and to encourage that loved one to continue to engage in activities that provide not only pleasure but a sense of fulfillment. A great suggestion is to encourage the loved one to check out Seniors Helping Seniors or enrolling the person with Alzheimer's in an adult day treatment program—anything that fills the void and leads to a more fulfilling life for the person with Alzheimer's disease.

In stage 5, moderate dementia, most people with Alzheimer's have a diagnosis and it is not a secret to anyone who knows this person. The individual is functioning at the level of a 5–7-year-old. The implications for care during this stage include assistance with bathing and dressing. The person may not be able to properly sequence dressing or make the appropriate choices for the occasion and/or for the weather. For example a woman might put her bra on over her blouse; a gentleman might put his pajama bottoms on to go to the doctor's office. At this stage, the person can still physically dress him/herself but would require some supervision to ensure that he/she was not dressing in a manner that would draw attention and lead to ridicule.

Sometimes people at stage 5 appear to live alone but benefit from daily oversight from either family members, neighbors, or members of their church or synagogue. On the surface, it might appear that this person is living independently but might be receiving assistance in purchasing and preparing food, making sure that rent and utilities are paid, and that financial matters are handled. This support from a distance is not ideal and certainly increases the safety risks for the person with Alzheimer's. This person is not independent in making doctor's appointments or arranging transportation. These tasks would need to be assumed by either a family member or a caring friend.

Early stage—abilities—"great foolers" Cognitive function 11–5 years of age	Early stage disabilities
Able to complete self-care tasks Able to communicate needs and converse Able to ask for help Able to structure ordinary daily routines	Forgetfulness/short-term memory loss may admit memory not good Impairment in judgment—makes bad decisions Difficulty with calculations, handling money Routine tasks take longer
Able to follow simple, verbal instructions Able to learn in situational-specific arenas—if skill is valued Able to understand and play familiar games Able to ambulate independently and safely if no physical impairment Able to use highly familiar tools safety—may need supervision for quality of work and for unforeseen hazards	Lack of safety awareness— poor understanding of physical deficits Difficulty with familiar tasks such as cooking, balancing the check book (easily scammed by dishonest people who prey on those with Alzheimer's), paying bills Difficulty finding specific words Lack of spontaneity Less initiative Trouble understanding long explanations, use of new devices, or secondary effects of action Becomes anxious easily or may have a tendency to withdraw I ncreasing disorientation regarding time and place—may begin to get lost driving a car Social withdrawal or depression Mood/personality changes

Stage 6 is the *moderately severe* stage according to Dr. Reisberg's FAST scale (Reisberg 1988) and the individual is functioning at the level of a toddler, 4 years of age deteriorating to 2 years of age. It is probably the most difficult stage for caregivers—lasting on an average 2–4 years but could last much longer. With each progressive step in the disease, the person with Alzheimer's loses additional skills. The individual is moving towards requiring total care. At the FAST level of 6a, the person will need assistance in choosing and putting on clothes in the correct sequence; at 6b the person will no longer be able to bathe or brush teeth independently but will require step-by-step assistance with these activities. At level 6c, the individuals will require assistance with the mechanics of toileting, i.e., flushing toilet paper down toilet rather than throwing it on the floor of the bathroom, pulling pants down so that urine and stool go into the toilet and not on the floor, in the trash can, or on the person's hands. At level 6d, urinary incontinence occurs followed by fecal incontinence follows at stage 6e. Frequent and scheduled toileting can be invaluable in managing the incontinence issues. Trips to the bathroom also need to be followed by scheduled drinks of water, juice, milk, or any other liquid. The strategy of a timed voiding schedule followed by 8 ounces of something to drink will go a long way towards not only managing the incontinence but also preventing the development of a urinary tract infection which is the number one cause of delirium in the elderly, specifically in those with Alzheimer's disease. It is so important to continually "go back" to where the person is cognitively and functionally. If we were taking care of a toddler between 2 and 4 years of age, we would ensure that the toddler received fluid all day long. We need to provide the same for those elderly who are functioning at the level of a toddler.

Stage 6 is also characterized by many challenging behaviors—resistance to bathing, incontinence, inability to dress self, aggression, screaming, wandering, repetition, sexually inappropriate behaviors, and more.

The following table represents a summary of the abilities and disabilities seen in stage 6 of the FAST scale.

Stage 6—toddlers—abilities	Stage 6—toddlers—disabilities
2–4 years of age Lasts 2–8 years Able to initiate familiar activity if supplies are available and in reach Able to do steps of self-care with verbal and tactile cues Able to tell stories from past Able to read words slowly out loud Able to follow slow, simple instructions Able to speak in short sentences or phrases; able to make needs known Able to sort, stack objects and do repetitive behaviors Able to sing, move to music, count Able to ambulate if no physical disability Able to feel and name objects (a toddler in an adult body)	Needs full-time supervision Only 5 min of short -term memory Problems recognizing family and friends Problems organizing thoughts/ logical thinking Repeats statements and/or movements Trouble dressing—may not want to bathe Increasing disorientation and forgetfulness Cannot find words—unconsciously fills in the blanks Suspicious, teary, fidgety, irritable, silly Challenging behaviors apparent

Stage 7 on the FAST describes the deficits in late dementia. It represents the end stage where the cognitive ability of the person with Alzheimer's 18 months deteriorating to that of a newborn.

Third stage (final) abilities Cognitive ability 18 months—newborn Lasts 1–3 years Abilities Appropriate for hospice care when stage 7C on FAST Complete dependence Can smile at the beginning of the end stage Can swallow thickened liquids at the beginning of the end stage May put everything in mouth or touch everything	Third (final) stage (1–3 years) Cognitive ability 18 months—newborn Lasts 1–3 years Disabilities Appropriate for hospice care when stage 7C on FAST 7a. Ability to speak limited (1–5 words a day) 7b. All intelligible vocabulary lost 7c. Loses ability to walk and becomes bedridden *Eligible for hospice at stage 7c based on the diagnosis of Alzheimer's disease 7d. Unable to sit up independently 7e. Loses the ability to smile 7f. Unable to hold head up

In Chap. 2, we review the areas of the brain that are damaged by Alzheimer's and the challenging behaviors that result from that damage.

References

Alzheimer's Association. (2014). Alzheimer's disease facts and figures. http://www.alz.org/downloads/Facts_Figures_2014. *Alzheimer's and Dementia, 10*(2), 16.

Aristophanes. (1938). The clouds (Anonymous translation). In W. J. Oates & E. J. O'Neil (Eds.), *The complete Greek drama* (p. 595). New York: Randme House. (original work published in 423 BC).

Auer, S. R., Sclan, S. G., Yaffee, R. A., & Reisberg, B. V. (1994). The neglected half of Alzheimer disease: Cognitive and functional concomitants of severe dementia. *Journal of the American Geriatrics Society, 42,* 1266–1272.

Bobinski, M., Wegiel, J., Wisniewski, H. M., Tarnawski, M., Bobinski, M., Reisberg, B., de Leon, M. J., & Miller, D. C. (1996). Neurofibrillary pathology-correlation with hippocampal formation atrophy in Alzheimer disease. *Neurobiology of Aging, 17,* 909–919.

Franssen, E. H., Kluger, A., Torossian, C. L., & Reisberg, B. (1993). The neurologic syndrome of severe Alzheimer's disease: Relationship to functional decline. *Archives of Neurology, 50,* 1029–1039.

Latham, A. (Ed.). (1967). *As you like it*. London: Methuen.

NIH Senior Health Gov. (2012). NIH Senior Health.Gov. http://nihseniorhealth.gov/alzheimersdisease/whatisalzheimersdisease/01.html.

Penny Wise, Pound Foolish: Fairness and Funding at the National Institute on Aging. (2011). Penny Wise, Pound Foolish: Fairness and Funding at the National Institute on Aging. www.alzfdn.org/documents/NIA%20Report-Final.pdf. Accessed 24 Mar 2014.

Prichep, L. S., John, E. R., Ferris, S. H., Rausch, L., Fang, Z., Cancro, R., Torossian, C., & Reisberg, B. (2006). Prediction of longitudinal cognitive decline in normal elderly with subjective complaints using electrophysiological imaging. *Neurobiology of Aging, 27,* 471–481.

Reisberg, B. (1988). *Functional assessment staging. Psychopharmacology Bulletin, 24,* 653–659.

Reisberg, B., & Franssen, E. H. (1999), Clinical stages of Alzheimer's disease. In M. J. deLeon (Ed.), *The encyclopedia of visual medicine series: An atlas of Alzheimer's disease* (vol. 139, pp. 1136–1139). Carnforth: Parthenon.

Reisberg, B., Ferris, S. H., de Leon, M. J., & Crook, T. (1982). The global deterioration scale for assessment of primary degenerative dementia. *American Journal of Psychiatry, 139*(9), 1136–1139.

Reisberg, B., Franssen, E. H., Souren, L. E. M., Auer, S. R., Akram, I., & Kenowsky, S. (2002). Evidence and mechanisms of retrogenesis in Alzheimer's and other dementias: Management and treatment import. *American Journal of Alzheimer's Disease, 17,* 202–212. (PubMed#12184509).

Reisberg, B., Shulman, M. B., Torossian, C., Leng, L., & Zhu, W. (2010). Outcome over seven years of healthy adults with and without subjective cognitive impairment. Alzheimer's & Dementia: *The Journal of the Alzheimer's, 6*(1), 11–24.

Chapter 2
There Is No Such Thing As a "Little Dementia"!

Families frequently report that they have received the diagnosis of "a little dementia" for a loved one whose memory is impaired. There is no such thing as "a little dementia." It is like being a "little pregnant"—either you have it or you do not! And having "dementia" is never a good thing—although families who receive that diagnosis will frequently say, "Well at least it is not Alzheimer's!" This denial keeps them from learning about the disease, planning for a time when the loved one with the diagnosis is unable to survive without 24 hour care, and making other decisions that are best made before a situation becomes a crisis.

The words of the physician carry incredible weight with families. It is important for families to hear that the physician suspects that their loved one has Alzheimer's dementia and not "a little dementia." Minimizing the seriousness of Alzheimer's is a great disservice not only to the primary caregiver but also to the family. For many people "a little dementia" does not sound very serious and certainly not life changing. But Alzheimer's disease is both serious and life changing not only for the person with the diagnosis but also for the primary caregiver to the person whose memory is failing, as well as to other members of the family. The reality is that Alzheimer's disease always ends in death. If the person progresses through all the stages of Alzheimer's, the disease is fatal. There is nothing about it that justifies a description of "a little dementia." It is a terrible disease of the brain that leads to the complete loss of all the functions that we associate with being an independent adult. An honest diagnosis allows families to plan for the future, to get their financial affairs in order, to seek out sources of education and support, to know what resources exist in her/his community and how to access them, and to learn about the disease and strategies to manage the inevitable progression and decline (Carson 2011).

In the newly released *DSM-5,* physicians are instructed, for all neurocognitive conditions to specify whether the condition is due to Alzheimer' disease, frontotemporal lobar degeneration, Lewy body disease, or a variety of other brain conditions

V. Benner Carson et al., *Care Giving for Alzheimer's Disease,*
DOI 10.1007/978-1-4939-2407-3_2

(American Psychiatric Association 2013). Are there other dementias that might also be linked to challenging behaviors? Of course there are and a brief description of selected dementias follows. But it is important to keep in mind that Alzheimer's dementia makes up approximately 65 % of all the dementias (Table 2.1).

Table 2.1 Common types of dementia

Alzheimer's disease (AD)	An estimated 5.2 million Americans have AD—including 5 million aged 65 and older or one in nine people aged 65 and older. Of those aged 75 or older, 44 % have it. One third (32 %) of those aged 85 and older have AD. Approximately 200,000 under age 65 have early onset AD
Vascular dementia (VD)	VD is the second most common cause of dementia after Alzheimer's disease, and is caused by "mini strokes" or occlusions of blood vessels in the brain. VD "paves the way" for Alzheimer's disease to develop, and VD/ALZ is a frequent diagnosis
Dementia with Lewy bodies (DLB)	People with Lewy body dementia often have memory loss and thinking problems similar to that seen in AD. However, the initial presentation of DLB includes sleep disturbances, hallucinations, and muscle rigidity or other movements that mimic Parkinson's disease. The brain changes of DLB alone can cause dementia, or these brain changes can coexist with the brain changes of Alzheimer's disease and/or vascular dementia, with each abnormality contributing to the development of dementia. When this happens, the individual is said to have "mixed dementia"
Parkinson's disease (PD)	As PD progresses, it often results in a progressive dementia similar to DLB or AD. Symptoms include problems with movement early on in the disease. If dementia develops, symptoms are often similar to DLB. The brain changes involve alpha-synuclein clumps that develop deep in the brain in an area called the substantia nigra. These clumps are thought to cause degeneration of the nerve cells that produce dopamine
Frontotemporal dementia (FTD)	Typical symptoms of FTD include changes in personality and behavior and difficulty with language. Nerve cells in the front and side regions of the brain are especially affected. Some people with FTD lose empathy for others as well as a sense of "what is appropriate" to say and do in public settings. Others lose language skills. There are no distinguishing microscopic abnormalities linked to all cases. People with FTD generally develop symptoms at a younger age (at about age 60) and survive for fewer years than those with AD

Table 2.1 (continued)

Creutzfeldt–Jakob disease (CJD)	Prion diseases such as CJD occur when prion protein, which is found throughout the body but whose normal function is not yet known, begins folding into an abnormal three-dimensional shape. CJD develops when prion protein in the brain also begins to fold into the same abnormal shape Through a process scientists do not yet understand, misfolded prion protein destroys brain cells. Resulting damage leads to rapid decline in thinking and reasoning as well as involuntary muscle movements, confusion, difficulty walking, and mood changes. Sign up for our enews to receive updates about Alzheimer's and dementia care and research CJD is rare, occurring in about one in 1 million people annually worldwide Experts generally recognize the following main types of CJD: Sporadic CJD develops spontaneously for no known reason. It accounts for 85 % of cases. On average, sporadic CJD first appears between ages 60 and 65 Familial CJD is a heredity form caused by certain changes in the prion protein gene. These genetic changes are "dominant," meaning that anyone who inherits a CJD gene from an affected parent will also develop the disorder. Familial CJD accounts for about 10–15 % of cases Infectious CJD is an especially rare form of CJD and results from exposure to an external source of abnormal prion protein. These sources are estimated to account for about 1 % of CJD cases. The two most common outside sources are: Medical procedures involving instruments used in neurosurgery, growth hormone from human sources or certain transplanted human tissues. The risk of CJD from medical procedures has been greatly reduced by improved sterilization techniques, new single-use instruments and synthetic sources of growth hormone The brain's patterns of electrical activity is similar to the way an electrocardiogram (ECG) measures the heart's electrical activity Brain magnetic resonance imaging (MRI) can detect certain brain changes consistent with CJD Lumbar puncture (spinal tap) tests spinal fluid for the presence of certain proteins Causes and risks Sporadic Creutzfeldt–Jakob disease has no known cause. Most scientists believe the disease begins when prion protein somewhere in the brain spontaneously misfolds, triggering a "domino effect" that misfolds prion protein throughout the brain. Genetic variation in the prion protein gene may affect risk of this spontaneous misfolding Mutations in the prion protein gene also may play a yet-to-be-determined role in making people susceptible to infectious CJD from external sources. Scientists do not yet know why infectious CJD seems to be transmitted through such a limited number of external sources. Researchers have found no evidence that the abnormal protein is commonly transmitted through sexual activity or blood transfusions

	Familial CJD is caused by variations in the prion protein gene that guarantee an individual will develop CJD. Researchers have identified more than 50 prion protein mutations in those with inherited CJD. Genetic testing can determine whether family members at risk have inherited a CJD-causing mutation. Experts strongly recommend professional genetic counseling both before and after genetic testing for hereditary CJD Sign up for our weekly e-newsletter Stay up-to-date on the latest advances in Alzheimer's and dementia treatments, care and research. Subscribe now Treatment and outcomes There is no treatment that can slow or stop the underlying brain cell destruction caused by Creutzfeldt–Jakob disease and other prion diseases. Various drugs have been tested but have not shown any benefit. Clinical studies of potential CJD treatments are complicated by the rarity of the disease and its rapid progression Current therapies focus on treating symptoms and on supporting individuals and families coping with CJD. Doctors may prescribe painkillers such as opiates to treat pain if it occurs. Muscle stiffness and twitching may be treated with muscle-relaxing medications or antiseizure drugs. In the later stages of the disease, individuals with CJD become completely dependent on others for their daily needs and comfort CJD progresses rapidly. Those affected lose their ability to move or speak and require full-time care to meet their daily needs. An estimated 90 % of those diagnosed with sporadic CJD die within one year. Those affected by familial CJD tend to develop the disorder at an earlier age and survive somewhat longer than those with the sporadic form, as do those diagnosed with vCJD. Scientists have not yet learned the reason for these differences in survival Currently, the only treatment is supportive—there is no cure. Doctors may prescribe painkillers such as opiates to treat pain if it occurs. Muscle stiffness and twitching may be treated with muscle-relaxing medications or antiseizure drugs. In the later stages of the disease, individuals with CJD become completely dependent on others for their daily needs. The progression of the disease is rapid and results in the loss of the ability to speak or move, requiring complete full-time care
Normal pressure hydrocephalus (NPH)	Symptoms of NPH include difficulty walking, memory loss, and inability to control urination, which are caused by the buildup of fluid in the brain. NPH can sometimes be corrected with surgery
Huntington's disease dementia (HDD)	HDD is a progressive brain disorder caused by a single defective gene on chromosome 4. Symptoms include abnormal involuntary movements, a severe decline in thinking and reasoning skills, and mood changes including irritability, depression, and others. The gene defect causes abnormalities in a brain protein that, over time, lead to worsening symptoms

Challenging Behaviors: Blame the Brain!

Another area that calls for honest discussion is the connection between challenging behaviors and specific brain damage. Why? Because without this knowledge, caregivers often arrive at the conclusion that their loved one with Alzheimer's is just being "difficult" and deliberately so! It is much more challenging to be patient with someone who seems to have some control over his/her behavior and is choosing to act in an unkind and difficult manner than it is to be patient with someone whose behaviors are directly linked to brain damage. Knowing that the behaviors result from brain damage removes personal responsibility from the patient—the behaviors naturally flow from what is happening in the brain and the behaviors are not deliberate attempts to frustrate or "get even with" the caregiver (Swaab 2014, pp. 348–351). Armed with knowledge connecting specific brain damage to specific behaviors allows caregivers to learn strategies to respond to these behaviors. It is not necessary for caregivers to have in-depth knowledge about the brain. A superficial, yet specific knowledge of how damage is linked to challenging behaviors is enough. Let us take a look at what level of knowledge is useful to caregivers. (Carson and Smarr 2007)

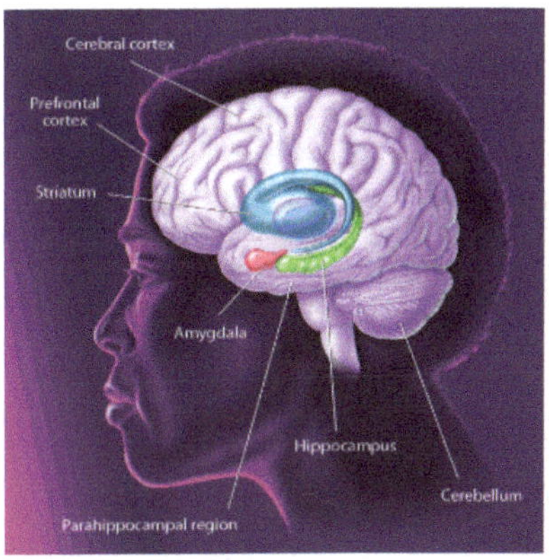

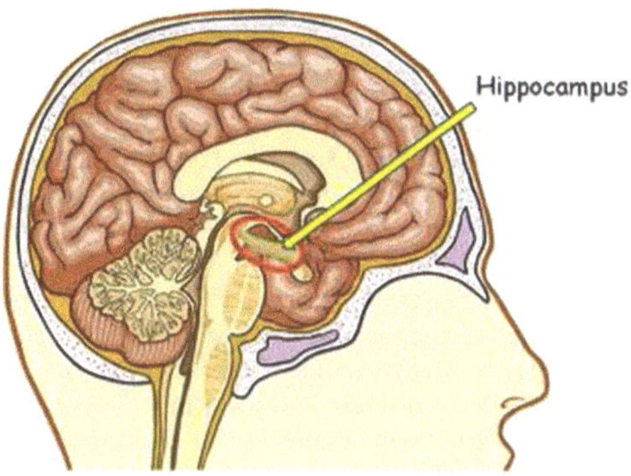

An area of the brain that is damaged very early on in Alzheimer's is the hippocampus shown in the picture on the left. This structure processes every experience that we have and stores long-term memories—sending the short-term memories to another part of the brain. In Alzheimer's, the brain becomes less and less able to process short-term memories, and since the "old" memories still remain, the person begins to "live" increasingly "in the past."

By stage 6 on the functional assessment scale (FAST) scale, the person has no more than 5 min of short-term memory and can only participate in one activity at a time. The lack of short-term memory means that questions are repeatedly asked and stories are retold. This is because the person asking the questions and retelling the stories has no memory of having repeatedly asked the questions or told the same story. How can a caregiver calmly respond to this repetition? There are many ways that caregivers can make use of the person's deficits. If the activity is repetitive, mindless, and yet productive, the caregiver might redirect the person to participate in repetitive activities that he or she enjoys.

Folding towels or other laundry is an activity that most women have done repeatedly throughout their lives. It is mindless, repetitive, and productive. Asking a woman to fold laundry is a strategy that makes her feel productive and useful while at the same time redirecting her repetitive questions. And just as important, if the person is not asking the same question repeatedly, the caregiver will be more likely to remain calm and loving. There are many other activities that can serve to redirect annoying repetition into useful tasks—it only takes a little imagination, and the ability to recognize that disabilities can many times be abilities—given the right situation.

In contrast, a man who has Alzheimer's and engages in repetitive questions will most likely not want to fold laundry. However, he could be given a container of mixed coins and be asked to separate the coins into separate piles to be rolled and taken to the bank. He might manipulate Legos® and sort them by color or design and might even build with them. Keep in mind that the use of Legos® does not imply that the caregiver is treating this older adult as a child. If the caregiver knows the person's "story" then the interventions can be individually tailored to that person's interests. An example will clarify this.

> An elderly man was living with his son and daughter-in-law and they were both concerned over what they saw as Dad's failing memory, his repetition and other behaviors that were troublesome. They insisted that dad see a physician who specialized in geriatric medicine. That physician did a complete workup of this gentleman and concluded that the man was in the early stages of Alzheimer's, stage 4 on the FAST scale. The gentleman had been a wood carver since he was a young man. His carvings were quite beautiful and people used to pay a good deal of money to purchase one of his carvings. When the son and his wife received the diagnosis that Dad had Alzheimer's disease, one of their first questions to the doctor was, "Should we take away Dad's knives?" The physician answered them with an emphatic "no." He told them that the carving was second nature to their father and he would most likely continue to be safe for quite some time. He said "just watch him—you will know when it is time to take away his carving knives." Several years passed before the son and his wife needed to place Dad in an assisted living facility. They still did not stop his carving. His daughter-in-law kept him supplied with bars of Ivory soap and plastic knives and the gentleman continued to lean forward in a chair with a trash can between his knees and carve the bars of soap.

The hippocampus along with the parietal lobe is critical to mapping skills and safely traveling from one location to another. These areas of the brain sustain damage early on in AD and are reflected in people getting lost—not only in areas many miles from home but also in areas close to home where the person would be expected to "know his way." Becoming lost is a common and dangerous behavior frequently seen in those with Alzheimer's disease (Chiu et al. 2004). For example, the following story was reported in the periodical *Alzheimer's Reading Room* in 2012. This potentially horrific story had a happy ending because two dedicated and kind police officers went "above and beyond the call of duty" (http://www.alzheimersreadingroom.com/2012/02/alzheimers-patient-lost-wanders-1500.html). Here is the cliff notes version of this story:

- *An unnamed man suffering from Alzheimer's managed to get on a bus in Virginia and traveled to Denver, Colorado.*
- *Someone discovers the man who at this point is disoriented, out of cash, and unable to cash a check.*
- *Next, police officer Hana Ruiz comes on to the scene.*
- *Police finally figure out his name by referring to his checkbook and are then able to contact his caregiver through the bank.*
- *Money is wired to buy him a plane ticket home.*
- *Happy ending right? Not yet.*

Next:

- *Officer Ruiz dips into her own pocket to buy him food, and gets him a place to stay.*
- *The next morning, Officer Ruiz asks Officer Rob Martinez to help get the man dressed and to the airport on time. Understandable.*
- *Officer Martinez notices that the man's cloths are in poor condition so he buys him new clothes. Officer Martinez also dips into his own pocket to do this.*
- *They take the man to the airport.*
- *Happy ending right?*
- *Not exactly, the man for some unknown reasons cannot get on the plane.*

Next:

- *Officer Ruiz dips into her own pocket to buy the man a bus ticket home.*
- *She then finds someone to accompany him home. Whew, great idea. If not, this story would probably have a part 2, 3, and 4.*
- *Finally, the two officers use their own money to make sure the man has food for the long trip home; and then, assured that his caseworker was waiting to pick him up in Virginia.*

Officers Ruiz and Martinez received the "Citizens Appreciate Police Award" (DeMarco 2012).

Because "getting lost" is such a common behavior among those with Alzheimer's, caregivers are encouraged to enroll their loved one in the Medic Alert + Safe Return program offered by the Alzheimer's Association. This program sells a *Medic-Alert* bracelet for the person with Alzheimer's as well as for the primary caregiver. If the person with Alzheimer's or a related dementia wanders and gets lost, caregivers can call an emergency response line (1-800-625-3780) and report it. A community support network is activated including law enforcement and members of the local Alzheimer's Association Chapter to assist in the search for the missing individual. If the person with Alzheimer's is found by someone other than the family or the police, he/she can call the toll free number listed on the back of the bracelet and Medic Alert + Safe Return will notify the person's contacts and make sure that the person is safely returned home. The necessity for the Medic Alert bracelet for the primary caregiver is best illustrated with another story.

> A daughter was caring for her father; they lived in a small suburban community in Michigan. One morning the daughter realized she needed to purchase a gallon of milk. She decided it was easier to go to the local super market without her father. Unfortunately she did not anticipate that she would be involved in a serious accident that would result in her hospitalization. Her car was struck by a large truck and she sustained life threatening injuries. Fortunately she was wearing her Medic Alert bracelet that identified her as a caregiver to someone with Alzheimer's disease. The hospital staff dispatched an ambulance to go to the daughter's home and pick up her father who was temporarily placed in an Assisted Care Facility until his daughter fully recovered from her injuries. (Story shared with Dr. Carson 2008)

Today, there are a variety of devices that utilize GPS technology so that the movements of the person with Alzheimer's can be tracked. Some states issue "Silver Alerts" which serve as a public notification system that broadcasts information about missing persons—especially seniors with Alzheimer's disease, other types of dementia or other mental disabilities—in order to aid in their safe return.

Silver Alerts use a wide array of media outlets—such as commercial radio stations, television stations, and cable TV—to broadcast information about missing persons. Silver Alerts also use variable-message signs on roadways to alert motorists to be on the lookout for missing seniors. In cases in which a missing person is believed to have gone missing on foot, Silver Alerts have used reverse 911 systems to notify nearby residents of the neighborhood surrounding the missing person's last known location.

The activation criteria for the Silver Alert system vary from state to state. Some states limit Silver Alerts to persons over the age of 65, who have been medically diagnosed with Alzheimer's disease, dementia, or similar mental disability. Other states expand Silver Alert to include all adults with mental or developmental disabilities. In general, the decision to issue a Silver Alert is made by the law enforcement agency investigating the report of a missing person. Public information in a Silver Alert usually consists of the name and description of the missing person and a description of the missing person's vehicle and license plate number.

Return to the diagram of the brain and look at the diagram on the left. The hypothalamus is located in close proximity to the hippocampus. The hypothalamus controls appetite and body temperature. At the beginning of stage 6 on the FAST scale, the person is hungry all the time, despite the size of the most recent meal! The hypothalamus stops sending signals of "fullness" even after eating a large meal. The person not only feels hungry but because of damage to the hippocampus the person may have no memory of recently having finished eating. The person will repeat "When are we going to eat?" over and over again. The caregiver needs to be instructed to make available healthy finger foods such as cheeses, fruit, vegetables, and pieces of meat and to encourage the person with Alzheimer's to graze throughout the day. If the person gains weight this is not a problem. If the person has diabetes and there is concern about blood sugar levels, this can also be managed. The reason is that as the person reaches the end of stage 6, the hypothalamus will stop sending signals of hunger so that the person will not want to eat and can lose as much as 20–30 % of their total body weight. A little extra weight will be a good thing when the person reaches the stage when refusal to eat is common. One of the strategies to encourage continued eating in the end stage is to puree food mixed with chocolate syrup or jelly, since many people with Alzheimer's retain their taste for sweets.

Let us look at what happens to body temperature in someone who has Alzheimer's disease. People with this disease are cold all the time—even when the ambient temperature is set high. The person might be wearing layers of clothing, knit booties on his/her feet, covered with a lap blanket and still complaining vociferously about the cold temperature. This has significant implications for bathing. The person who will be bathing another person needs to build up a head of steam in the bathroom so it is warm in there, and towels and clean clothes need to be warmed before offering them to the person.

What about the limbic system, the center for emotional control? As the limbic system sustains increasing damage from Alzheimer's, the person will gradually show symptoms of being on an emotional roller coaster. She is happy 1 min and then without any apparent external provocation she is angry or teary eyed! The response of the caregiver needs to be comforting and patient towards the loved one. However, the patient does not "own" this behavior. Without knowledge that this rapid change in moods is caused by brain damage, the caregiver might be very perplexed to understand these emotional changes and might assume unnecessary guilt for inducing them.

What about the occipital lobe? Damage to the occipital lobe can result in visual agnosia. In other words, people see an object but their brains no longer can interpret the purpose of the object. For instance, a person with Alzheimer's might pick up a toothbrush and try to brush his/her hair with it. Or, a person might think that the television remote control is a telephone receiver. It is not that the person cannot "see" the object, but that his/her brain can no longer identify its use.

The last area of the brain that can be damaged by Alzheimer's is the frontal lobe which can lead to behavioral dysregulation resulting in behaviors such as saying or doing things that might be considered threatening, bizarre, or generally inappropriate. Examples of these behaviors might include swearing, undressing, urinating in public, eating and drinking nonfood items and others—all very difficult for caregivers to manage. The frontal lobes are essential to planning, organizing, and creating structure in our days—these are skills that are lost in Alzheimer's.

In the next chapter, we will take a look at pain—a condition that is largely ignored in those with Alzheimer's disease. Because pain is frequently not recognized, those with Alzheimer's may receive antipsychotic medications that may or may not be appropriate.

References

American Psychiatric Association. (2013). *Diagnostic and statistical manual of mental disorders* (5th ed.). Arlington: American Psychiatric.

Carson, V. B. (2011). Dementia or Alzheimer's: What is the difference? *Caring Magazine, 30*(7), 40–41.

Carson, V. B., & Smarr, R. (2007). Becoming an Alzheimer's whisperer. *Home Healthcare Nurse, 25*(10), 628–636.

Chiu, Y. C., Algase, D., Whall, A. (2004). Getting lost: Directed attention and executive functions in early Alzheimer's disease patients. *Dementia Geriatric Cognitive Disorders, 17*(3), 174–180. (Accessed 4 April2014).

DeMarco, B. (2012). Alzheimer's patient lost, wanders 1500 miles from home. Alzheimer's Reading Room. (February 1, 2012). http://www.alzheimersreadingroom.com/2012/02/alzheimers-patient-lost-. Accessed 2 April 2014.

MedicAlert Foundation. http://www.medicalalert.org

Morris, J. C., Barrett, L. F., Dickerson. B. C. (2011). Amygdala atrophy is prominent in early Alzheimer's disease and relates to symptom severity. *Psychiatry Research: Neuroimaging, 194*(1), 7–13, 31.

Swaab, D. F. (2014). *We are our brains: A neurobiography or the brain, from the womb to Alzheimer's*. New York: Spiegel & Grau.

Chapter 3
"If I knew My Loved One Was in Pain…"

Pain is often unrecognized in the elderly with Alzheimer's disease (AD). As a result, challenging behaviors such as resistance to care, hitting, screaming, biting, refusal to eat, and others are attributed to deliberate resistance by the person with Alzheimer's, or it may be assumed that the disease has led to psychotic behavior which would warrant the use of antipsychotic drugs. However, a clinical assessment for pain is usually absent. Failure to recognize pain, then, often leads to overmedicating with powerful antipsychotic and antianxiety medications with all their attendant side effects. Let us take a look at a case study of a woman who was recently admitted to an assisted living facility.

> This story was shared by Mary, a home health care nurse.
> I was referred a patient, Mrs. S who was an 81-year-old female diagnosed with AD. Mrs. S had been living in an assisted living facility for the past month and a half. She had exhibited escalating behaviors where she threw objects at staff members, tried to escape from the memory care facility, displayed constant agitation, continued crying, and hits her head with a clenched fist. She had also hit a staff member who attempted to get her up out of a chair. When I first met her, she was heavily sedated with Depakote and Ativan. These drugs were prescribed by a house physician who was contacted by a facility nurse after the patient hit a staff member. The patient's Depakote had been on hold for the past 2 days and today (second attempt at a visit) was the first time in 5 days that she opened her eyes. She had been refusing to open her mouth to put her dentures in and therefore she had also refused to eat. She was unable to ambulate. She was also on antibiotics for a current urinary tract infection.
> As part of my initial evaluation of Mrs. S., I met with her son and the Director of the assisted living facility. Mrs. S's son was questioned about whether Mrs. S had any conditions that would cause her to be in pain. He said that his mother used to have a lot of back pain as well as arthritic pain in her knees and hips. When she was at home living on her own, she took Tylenol® regularly and occasionally an Aleve® on really bad days. Since being admitted to the facility, she had not been given any Tylenol®. When she entered the facility, her son had only provided the staff with a list of her prescription meds, and no one had asked if Mrs. S. had any history of pain or was regularly taking any over-the-counter medications.
> After Mary completed the initial evaluation of Mrs. S, she spoke with Mrs. S.'s physician about reducing the more sedating medication and trying pain medication instead based on the results of the pain assessment that Mary had completed. The physician agreed with

V. Benner Carson et al., *Care Giving for Alzheimer's Disease*,
DOI 10.1007/978-1-4939-2407-3_3

Mary and the Depakote was discontinued and Ativan was also gradually reduced. Mrs. Smith was started on Acetaminophen 500 mg × 2 PO every 4 h while awake. Facility staff were then educated on assessment of pain using PAINAD and Wong–Baker FACES® pain rating scale. The nurse taught the staff members that pain in patients with Alzheimer's is often exhibited in agitated behaviors. Over the next couple of weeks, the sedative medication was discontinued. The pain medication was changed to Vicodin × 1 every 4 h as needed, based on the staff's assessment. The patient was alert and during nurse's last visit was folding napkins over and over again and chatting to herself. Her aggression towards the staff had totally stopped. Nursing staff were able to walk her to the toilet. They could now put in her full dentures and feed her and she was eating well. Sometimes, she still cried but no longer hit her head and clenched her fists.

These changes took place over 6 weeks. The nurse spoke to Mrs. Smith's son every week and he was happy with how things were going with his mother.

In 2012, the (Centers for Medicare, and Medicaid Services (CMS) issued a warning to skilled nursing providers across the country that their reliance on powerful mind altering medications to control challenging behaviors was unacceptable and amounted to "chemical restraints"(http://blogs.lawyers.com/2013/09/antipsychotic-drugs-over-used-in-nursing-homes). The challenge to these agencies was clear—to find behavioral responses to challenging behaviors or be faced with a sharp decrease in reimbursement. Although the Medicare and Medicaid programs prohibit the use of chemical restraints, antipsychotic medications continue to be used sometimes as a substitute for good, individualized care. Many believe that the use of chemical restraints has increased as nursing homes hire fewer staff members to work with residents (J Am Geriatr Soc 62:454–461, 2014). Sadly, patients who continue to reside at home are also often prescribed these same medications. Family members, faced with challenging behaviors, plead with the primary care physician to "do something" to calm their loved one. Many times these pleas for help follow an aggressive outburst—frequently, but not always, occurring when the caregiver attempts to shower the person with Alzheimer's. Why would showering be an activity that results in aggression?

Let us paint a picture of what typically happens. The person with Alzheimer's is taken into a cold bathroom, stripped of his/her clothes, and forced into a shower to be bathed. His hypothalamus is often not working correctly so his internal thermostat is set at a very low temperature—he was cold before he was stripped! He has arthritis in multiple joints; he is frightened by the shower and he strikes out at the caregiver attempting to bathe him. The patient pushes the caregiver and the caregiver hits her head on the hard tile floor resulting in a gash on her forehead which requires treatment in the local emergency department. The primary care physician is called and the family caregiver implores the doctor to order medication that will calm the patient. So, an antipsychotic medication is prescribed. Now the patient is still in pain but lacks the ability to communicate his pain. He cannot even push away a caregiver who is bathing him in a bathroom that is too cold and in a shower with lukewarm water that makes him shiver! It sounds horrid! It is.

Now let us look at an alternative reality. The person receives Tylenol® 30 min prior to going into the bathroom. The Tylenol® takes the edge off of arthritic pain. A few minutes prior to helping this person into the bathroom, the caregiver turns on the hot water and builds a head of steam in the bathroom. The caregiver then puts towels into the clothes dryer so that the towels will be warm against the person's body. The caregiver carries with her a pocket full of Tootsie Roll Pops® which she unwraps to give to the person needing to be bathed. She has learned that he can only do one thing at a time so if he is sucking on a Tootsie Roll Pop® she can bathe him with no problems. She wraps him with the warmed towels and removes his clothes from underneath the towels—he is never totally undressed. She has his favorite religious music playing on the CD and she sings Amazing Grace as she proceeds to bathe him. Does this approach take planning and probably a little more time? Of course, but the bottom line is that no one gets hurt; the patient is bathed and dressed in clean clothes; his pain is controlled, and he leaves the bathroom smiling! (Carson 2011 April)

Dangers of Antipsychotics in the Elderly

Antipsychotic drugs pose significant risks to the elderly person. According to a CDC Nursing Home survey in 2004, 47.9 % of residents were taking nine or more medications and 37.1 % were taking between 5 and 8 medications (CDC 2004). A 2003 study found that more than half of community-dwelling older adults who were admitted to nursing homes receive psychotropic medication within 2 weeks of their admission (Voyer and Martin 2003). Usage of psychotropic medications is linked with certain medical conditions such as respiratory infections and strokes. In 2005, the US Food and Drug Administration(FDA) issued an advisory and subsequent black box warning regarding the risks of atypical antipsychotic use among elderly patients with dementia (Arch Intern Med. 170:96–103, 2012). "Black Box" warnings for antipsychotics stated that individuals diagnosed with dementia are at an increased risk of death (60–70 %) (increased from 2.6 % of those receiving placebo to 4.5 % in those taking antipsychotics[1]. Physicians are advised to discuss this risk with their patients (or their guardians) before prescribing these drugs. Still, nearly half of patients with dementia are prescribed antipsychotics. (http://blogs.lawyers.com/2013/09/antipsychotic-drugs-over-used-in-nursing-homes/). It is important to keep in mind that the same scenarios are frequently played out in private homes where the family caregiver is at his/her wits end to know how to respond to screaming, hitting, and other challenging behaviors.

Why are these drugs so heavily prescribed? The answer lies in the fact that behavioral and psychological symptoms of dementia (BPSD) are a common and difficult to manage set of symptoms that might include aggression, violence, or socially inappropriate behaviors that challenge the management skills of families, nurses, and direct care workers. By not medicating those with Alzheimer's with these powerful drugs, the work of providing care might be more rather than less time consuming, and would not be possible for many caregivers who would have to admit their loved one to a nursing home. Antipsychotics have a long history of being used to

[1] See FDA website: http://www.fda.gov/Drugs/DrugSafety/PostmarketDrugSafetyInformationforPatientsandProviders/ucm152291.htm. Here is the black box warning: Elderly patients with dementia-related psychosis treated with antipsychotic drugs are at an increased risk of death. Risperdal [or other antipsychotic] is not approved for use in patients with dementia-related psychosis.)

reduce the impact of BPSD on patients and caregivers. However, the evidence for the efficacy of these drugs is modest at best and the side effects are quite serious. In fact, many are packaged with a Black Box Warning that their use to control BPSD in residents of SNIF'S (skilled nursing facilities), could result in death. The warning from CMS was in response to growing concern that these drugs are being used to inappropriately sedate residents with dementia is also growing concern that antipsychotics are being used as a means of behavior control.

Antipsychotics should only be used in severe cases of BPSD when nonpharmacological interventions have failed, and then only for short periods with regular review. In 2012, the Improving Dementia Care Treatment for Older Adults Act was introduced in Congress. It addresses growing costs and concerns about the overuse of antipsychotics in nursing homes and other long-term care facilities. It did not pass the 112th Congress. The Centers for Medicare and Medicaid Services launched the "Partnership to Improve Dementia Care" initiative, which was designed to reduce the misuse of antipsychotics among nursing home residents by 15 % by the end of 2102. This deadline was not met, but officials were hopeful that it would be met by the end of 2013. Unfortunately, that deadline was not met either.

There is work to be done in skilled nursing facilities!

What does all of this have to do with pain? Perhaps the "behaviors" that are being targeted with antipsychotic medications reflect unmanaged pain rather than psychotic behaviors. Why would the person with Alzheimer's be in pain? There could be any number of diagnoses that could account for pain but most likely this elderly person has arthritis—a leading cause of functional decline in the elderly. Arthritis is painful. Many elderly people have arthritis in every joint—their ankles, knees, hips, fingers, and spines. Prior to developing AD, the person with arthritis may have managed her/his own pain with Tylenol® and on a rainy day might add an Aleve®. If she/he has mentioned arthritis to the primary care physician she/he might be lucky enough to be on an anti-inflammatory medication (if the older person does not have kidney disease or stomach problems). The patient does not talk about the arthritis and barely complains—why? Because there is nothing other than taking these drugs that is available! The person just tries to "put up with the pain."

What happens when the individual with Alzheimer's has deteriorated to stage 6 on the functional assessment staging tool (FAST) scale and functioning at the level of a toddler? At that point, the individual no longer has the ability to verbally communicate his/her pain. What if the family or paid caregiver does not know the individual's history of arthritis? This probably means that there will be no pain medication provided. Pain is an abstraction—and requires a higher level of functioning to be able to identify pain as such. So, the person who is functioning at a toddler level feels pain but lacks the cognitive ability to effectively communicate that pain. Because caregivers are not aware that the person is experiencing pain, there are no analgesics offered or given. If the person experiencing pain is moved too quickly and pain is increased, she/he may respond by striking out at the person who has caused or is causing pain. What happens next? Predictably the person with Alzheimer's receives a powerful antipsychotic medication for "aggression"! This only further impedes that person's ability to express pain (Selbaek et al. 2007).

Consider how we recognize pain in a toddler. We notice that the child's color is "not good," he/she is listless and has no appetite, maybe the child is crying and whining throughout the day and not sleeping well at night. Most if not all caregivers of toddlers would immediately recognize these symptoms as an "illness" of some kind. Following a night of taking care of this sick toddler, the parent might call the pediatrician or take the child into an urgent care facility to seek medical help for this child. At no time would the caregiver request a medication to "knock out" the child so that he/she would sleep and also allow the parent to sleep! Such a request would be shocking and might warrant "raised eyebrows" at the least and at the most a referral to "child protective services."

The point is that we recognize pain and illness in a child but we frequently fail to recognize pain in an elderly person with Alzheimer's who is functioning at the level of a child. Pain in the person with Alzheimer's is "acted out" through crying, biting, hitting, kicking, and other aggressive behaviors that are directed towards stopping an unknowing caregiver from increasing pain through bathing, toileting, assisting out of bed, and other activities that cause pain. If the caregiver is not sensitive to these signs of pain, the most likely intervention will be the use of antipsychotic medications! Imagine the horror of being trapped in pain, dulled by powerful medications, and unable to communicate!

Quantifying pain in someone who cannot report the pain presents a challenge to both professional caregivers as well as family caregivers. Nevertheless, there are many easy to use and interpret pain scales that are designed for use with the person who has AD. One of the easiest to use measures developed for children and currently used worldwide as an assessment tool to quantify pain across the age span that works perfectly with an adult who is functioning at the level of a young child. This tool is the Whaley and Wong FACES Pain scale. It is the norm today in the USA that when a patient registers for care in an emergency department, a nurse will give that person a copy of the FACES pain scale and ask the patient to identify which face most accurately communicates the degree of pain that the patient is experiencing.

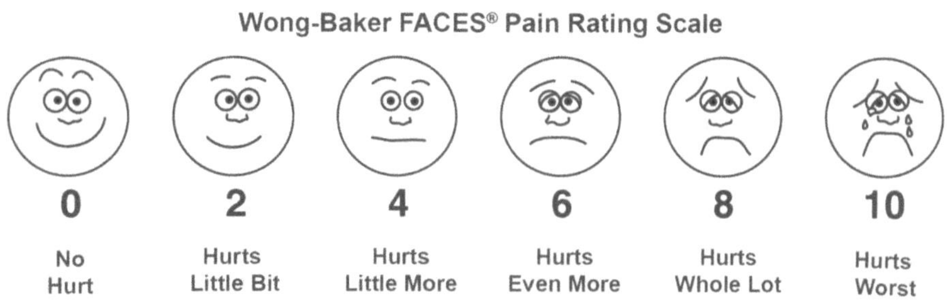

Instructions for Usage

Explain to the person that each face is for a person who has no pain (hurt), some pain, or a lot of pain.

Face 0 does not hurt at all. Face 2 hurts just a little bit. Face 4 hurts a little bit more. Face 6 hurts even more. Face 8 hurts a whole lot. Face 10 hurts as much as you can imagine, although you do not have to be crying to have this worst pain.

Ask the person to choose the face that best depicts the pain they are experiencing.

There are two other excellent tools used to assess and quantify pain in those *with* AD.

One is the Abbey Scale—developed in Australia (Abbey et al. 1998–2002).

Abbey Pain Scale
For measurement of pain in people with dementia who cannot verbalize

How to use scale: While observing the resident, score questions 1 to 6

Name of resident: ………………………………………………………………………………

Name and designation of person completing the scale: ………………………………

Date: ……………………………………………Time: ………………………………………

Latest pain relief given was…………………………………………………at ……….hrs.

Q1. Vocalization
eg. whimpering, groaning, crying
Absent 0 Mild 1 Moderate 2 Severe 3
Q1 ☐

Q2. Facial expression
eg: looking tense, frowning grimacing, looking frightened
Absent 0 Mild 1 Moderate 2 Severe 3
☐ Q2

Q3. Change in body language
eg: fidgeting, rocking, guarding part of body, withdrawn
Absent 0 Mild 1 Moderate 2 Severe 3
☐ Q3

Q4. Behavioural Change
eg: increased confusion, refusing to eat, alteration in usual patterns
Absent 0 Mild 1 Moderate 2 Severe 3
☐ Q4

Q5. Physiological change
eg: temperature, pu lse or blood pressure outside normal limits, perspiring, flushing or pallor
Absent 0 Mild 1 Moderate 2 Severe 3
☐ Q5

Q6. Physical changes
eg:skin tears, pressure areas, arthritis, contractures, previous injuries.
Absent 0 Mild 1 Moderate 2 Severe 3
Q6 ☐

Add scores for 1 6 and record here ⇨ Total Pain Score ☐

Now tick the box that matches the Total Pain Score ⇨

0 2 No pain	3 – 7 Mild	8 – 13 Moderate	14+ Severe

Finally, tick the box which matches the type of pain ⇨

Chronic	Acute	Acute on Chronic

Dementia Care Australia Pty Ltd
Website: www.dementiacareaustralia.com

Abbey, J; De Bellis, A; Piller, N; Esterman, A; Giles, L; Parker, D and Lowcay, B.
Funded by the JH & JD Gunn Medical Research Foundation 1998 2002
(This document may be reproduced with this acknowledgment retained)

The other is the *pain assessment in advanced dementia* (PAINAD). This tool evaluates the following five descriptive elements that are impacted by pain.

This pain tool is used to assess pain in older adults who have dementia or significant cognitive impairment that makes them unable to reliably communicate their pain. It can be used by a nurse or by a CNA to screen for pain-related behaviors. The tool should be used at admission, at scheduled quarterly assessments, during every shift if there is indication that pain is not controlled, each time a change in pain status is reported, and following a pain intervention to evaluate treatment effectiveness (within 1–2 h). The PAINAD is used by observing the older adult for 3–5 min during activity/with movement (such as bathing, turning, transferring). For each item included in the PAINAD, select the score (0, 1, 2) that reflects the current state of the behavior. Add the score for each item to achieve a total score. Total scores range from 0 to 10 (based on a scale of 0 to 2 for each of five items), with a higher score suggesting more severe pain (0 = "no pain" to 10 = "severe pain"). The score is reported by the nursing assistant to the nurse who does a follow-up assessment. The score is documented and shared with all other health care providers. Behavior observation scores should be considered alongside knowledge of existing painful conditions and reports from someone who knows the older adult (like a family member or nursing assistant) and their pain behaviors. It is very important to remember that some older adults may not demonstrate obvious pain behaviors or cues.

Pain Assessment in Advanced Dementia Scale (PAINAD)

Instructions: Observe the patient for 5 min before scoring his or her behaviors. Score the behaviors according to the following chart. Definitions of each item are provided on the following page. The patient can be observed under different conditions (e.g., at rest, during a pleasant activity, during caregiving, after the administration of pain medication) (Warden et al. 2003).

Behavior	0	1	2	Score
Breathing Independent of vocalization	Normal	Occasional labored breathing, short period of hyperventilation	Noisy labored breathing, long period of hyperventilation Cheyne–Stokes respirations	–
Negative vocalization	None	Occasional moan or groan Low-level speech with a negative or disapproving quality	Repeated troubled calling out Loud moaning or groaning Crying	–
Facial expression	Smiling or inexpressive	Sad Frightened Frown	Facial grimacing	–
Body language	Relaxed	Tense Distressed pacing Fidgeting	Rigid Fists clenched Knees pulled up Pulling or pushing away Striking out	– – –

Behavior	0	1	2	Score
Consolability	No need to console	Distracted or reassured by voice or touch	Unable to console, distract, or reassure	–
			Total score	–

Scoring: The total score ranges from 0 to 10 points. A possible interpretation of the scores is: 1–3 = mild pain; 4–6 = moderate pain; 7–10 = severe pain. These ranges are based on a standard 0–10 scale of pain, but have not been substantiated in the literature for this tool.

Source: Warden et al. (2003).

PAINAD Item Definitions

Source: Warden et al. (2003).

Breathing

1. *Normal breathing* is characterized by effortless, quiet, rhythmic (smooth) respirations.
2. *Occasional labored breathing* is characterized by episodic bursts of harsh, difficult, or wearing respirations.
3. *Short period of hyperventilation* is characterized by intervals of rapid, deep breaths lasting a short period of time.
4. *Noisy, labored breathing* is characterized by negative-sounding respirations on inspiration or expiration. They may be loud, gurgling, wheezing. They appear strenuous or wearing.
5. *Long period of hyperventilation* is characterized by an excessive rate and depth of respirations lasting a considerable time.
6. *Cheyne–Stokes respirations* are characterized by rhythmic waxing and waning of breathing from very deep to shallow respirations with periods of apnea (cessation of breathing).

Negative Vocalization

1. *None* is characterized by speech or vocalization that has a neutral or pleasant quality.
2. *Occasional moan or groan* is characterized by mournful or murmuring sounds, wails, or laments. Groaning is characterized by louder than usual inarticulate involuntary sounds, often abruptly beginning and ending.

3. *Low-level speech with a negative or disapproving quality* is characterized by muttering, mumbling, whining, grumbling, or swearing in a low volume with a complaining, sarcastic, or caustic tone.
4. *Repeated troubled calling out* is characterized by phrases or words being used over and over in a tone that suggests anxiety, uneasiness, or distress.
5. *Loud moaning or groaning* is characterized by mournful or murmuring sounds, wails, or laments in much louder than usual volume. Loud groaning is characterized by louder than usual inarticulate involuntary sounds, often abruptly beginning and ending.
6. *Crying* is characterized by an utterance of emotion accompanied by tears. There may be sobbing or quiet weeping.

Facial Expression

1. *Smiling or inexpressive.* Smiling is characterized by upturned corners of the mouth, brightening of the eyes, and a look of pleasure or contentment. Inexpressive refers to a neutral, at ease, relaxed, or blank look.
2. *Sad* is characterized by an unhappy, lonesome, sorrowful, or dejected look. There may be tears in the eyes.
3. *Frightened* is characterized by a look of fear, alarm, or heightened anxiety. Eyes appear wide open.
4. *Frown* is characterized by a downward turn of the corners of the mouth. Increased facial wrinkling in the forehead and around the mouth may appear.
5. *Facial grimacing* is characterized by a distorted, distressed look. The brow is more wrinkled, as is the area around the mouth. Eyes may be squeezed shut.

Body Language

1. *Relaxed* is characterized by a calm, restful, mellow appearance. The person seems to be taking it easy.
2. *Tense* is characterized by a strained, apprehensive, or worried appearance. The jaw may be clenched. (Exclude any contractures.)
3. *Distressed pacing* is characterized by activity that seems unsettled. There may be a fearful, worried, or disturbed element present. The rate may be faster or slower.
4. *Fidgeting* is characterized by restless movement. Squirming about or wiggling in the chair may occur. The person might be hitching a chair across the room. Repetitive touching, tugging, or rubbing body parts can also be observed.
5. *Rigid* is characterized by stiffening of the body. The arms and/or legs are tight and inflexible. The trunk may appear straight and unyielding. (Exclude any contractures.)
6. *Fists clenched* is characterized by tightly closed hands. They may be opened and closed repeatedly or held tightly shut.

7. *Knees pulled up* is characterized by flexing the legs and drawing the knees up toward the chest. An overall troubled appearance. (Exclude any contractures.)
8. *Pulling or pushing away* is characterized by resistiveness upon approach or to care. The person is trying to escape by yanking or wrenching him- or herself free or shoving you away.
9. *Striking out* is characterized by hitting, kicking, grabbing, punching, biting, or other form of personal assault.

Consolability

1. *No need to console* is characterized by a sense of well-being. The person appears content.
2. *Distracted or reassured by voice or touch* is characterized by a disruption in the behavior when the person is spoken to or touched. The behavior stops during the period of interaction, with no indication that the person is at all distressed.
3. *Unable to console, distract, or reassure* is characterized by the inability to soothe the person or stop a behavior with words or actions. No amount of comforting, verbal or physical, will alleviate the behavior.

Behavior observation scores should be considered alongside knowledge of existing painful conditions and reports from someone who knows the older adult (like a family member or nursing assistant) and their pain behaviors. It is so important to remember that some older adults may not demonstrate obvious pain behaviors or cues.

PAINAD score 0–2	Pain intensity No pain	Scores applied to: World Health Organization Analgesic Ladder approach
3–5	Mild to moderate pain	*STEP 1*: A non-opioid analgesic, i.e., acetaminophen or NSAID Adjuvant drugs if a specific condition exists (e.g., nerve pain)
6–8	Moderate to severe pain	*STEP 2*: An opioid conventionally used for moderate pain A non-opioid analgesic, i.e., acetaminophen or an NSAID An adjuvant drug in some cases Adjuvant drugs if a specific condition exists (e.g., nerve pain)
9–10	Severe pain	*STEP 3:* (Or for patients who failed to achieve adequate relief with step 2): An opioid used for severe pain A non-opioid analgesic in some cases (e.g., bone pain) Adjuvant drugs in some cases

Application to the World Health Organization's Analgesic Ladder done by Author: Verna Benner Carson, PhD, PMH/CNS-BC

End Stage and Pain

Care needs are extensive during the late stage and may be more than a family is able to provide at home. It is important for family caregivers to know that at stage 7C on the FAST scale, the person with Alzheimer's is eligible to receive hospice care either in that person's private home or within a facility. If the family chooses to provide that care at home, Medicare allows for respite care for the family. This may mean moving the person into a facility for a brief time in order for the family caregiver to get some respite. If the impending cause of death is a diagnosis other than Alzheimer's, e.g., metastatic cancer, the person requiring hospice could be at a much earlier stage on the FAST scale and be eligible for hospice care. Deciding on late-stage care can be one of the most difficult decisions that families face. It is important for families to think through this process prior to needing to make the decision so they do not second guess themselves after the fact. Ideally, discussions about end-of-life care wishes should take place while the person with the dementia still has the capacity to make decisions and share wishes about life-sustaining treatment.

One of challenges facing family as well as professional caregivers is managing pain. Effective pain management may have important implications for improving quality of life among individuals with end-stage Alzheimer's disease. It is important that all caregivers, family as well as professional, conduct an assessment of the person's pain level. Pain is often missed even when an Alzheimer's patient is in hospice care. Acetaminophen can be safely used around the clock in this population at total daily dose of up to 3000 mg (dropped from 4000 mg/day in 2011 due to liver toxicity). Although nonsteroidal anti-inflammatory drugs often work well for these patients, clinicians should be aware that they have various side effects (kidney, blood pressure, stomach). Additionally, it is important for clinicians to know that nonsteroidal anti-inflammatory drugs have a low ceiling effect (i.e., giving more of the drug above a certain dose does not result in more pain relief).

Opiate drugs, such as morphine, have no ceiling effect and have been shown to relieve all types of pain. Also, opiate drugs are probably underused in the geriatric population and may be safer than other drug strategies that are employed. However, clinicians should be aware that opiate drugs can increase confusion among patients with end-stage Alzheimer's disease, increase constipation, and increase risk of falls and hip fracture.

The transdermal fentanyl patch may be helpful among patients who are unable to swallow pills. However, because of the drug's extreme potency and the potential for overdose, it should not be used among elderly patients who are opiate naïve (have not taken opiate pain medication before) or among those who are unaccustomed to the respiratory depression caused by opiates. Finally, a nonnarcotic agent, tramadol, may be useful. Tramadol has the same potency as codeine, but it rarely causes respiratory depression (although tramadol is not considered a narcotic, it is an opioid drug similar to morphine, and while it is not as addicting as other opioids, can still be addictive and has many interactions with other medications—especially antidepressants; the risk of seizures is also increased with the use of this medication, as well as risk of serotonin syndrome when used with antidepressants).

References

Abbey, J., DeBellis, A., Piller, N., Esterman, A., Giles, L., Parker, D., & Lowcy, B. (1998–2002). Abbey Pain Scale, Funded by J.H. and Ginn Medical Research Foundation.

Abbey, J. A., Piller, N., DeBellis, A., Esterman, A., Parker, D., Giles, L., & Lowcay, B. (2004). The Abbey Pain Scale. A 1-minute numerical indicator for people with late-stage dementia. *International Journal of Palliative Nursing, 10*(1), 6–13.

Allen, M. (2014). One third of skilled nursing patients harmed in treatment, ProPublica, March 3rd 2014.

Arenella, C. (2007). Alzheimer's disease (and other brain diseases) hospice care. Hospice foundation. www.americanhospice.org. Accessed 4 April 2014.

Bonner, A. (2013). Improving dementia care and reducing unnecessary use of antipsychotic medications in nursing homes. Medicare learning network, Washington D.C. (January 1, 2013).

Carson, V. B. (2011). Taking the angst out of bathing. *Caring, 30,* 40–42.

Declercq, T., Petrovic, M., Azermai, M., Vander, S. R., De Sutter, A. I. M., van Driel, M. L., Christiaens, T. (2013). Withdrawal versus continuation of chronic antipsychotic drugs for behavioural and psychological symptoms in older people with dementia. The Cochrane Collaboration®. Publsihed by John Wilet & Sons, Ltd.

Ferrell, B. A. (1995). Pain evaluation and management in the nursing home. *Annals of Internal Medicine, 123,* 681–687.

Ferrell, B. A. (1996). *Pain evaluation and management, in Medical Care of the Nursing Home Resident*. Philadelphia: American College of Physicians. (Edited by Besdine RW, Rubenstein LZ, Snyder L).

Ferrell, B. A. (2004). Managing pain and discomfort in older adults near the end of life. *Annals of Long-Term Care, 12,* 49–55.

Ferrell, B. A., Ferrell, B. R., Rivera, L. (1995). Pain in cognitively impaired nursing home patients. *Journal of Pain and Symptom Management, 10,* 591–598.

Konetzka, R., Tamara, D. J., Brauner, J. S., & Werner, R. M. (2014). The effects of public reporting on physical restraints and antipsychotic use in nursing home residents with severe cognitive impairment. *Journal of the American Geriatrics Society, 62,* 454–461.

Moss, M. S., Braunschweig, H., Rubinstein, R. L. (2002). Terminal care for nursing home residents with dementia. *Alzheimer's Care Quarterly, 3,* 233–246.

Portenoy, R. K. (1999). Opiate therapy for chronic non-cancer pain: can we get past the bias? *American Pain Society Bulletin, 1,* 4–7.

Selbaek, G., Kirkevold, O., Engedal, K. (2007) The prevalence of psychiatric symptoms and behavioural disturbances and the use of psychotropic drugs in Norwegian nursing homes. *International Journal of Geriatric Psychiatry, 22,* 843–849.

Singer, P.A., Martin, D.K., & Comment, M. (1999). Quality end-of-life care: Patients' perspectives. *Journal of American Medical Association, 281*(16), 1488.

Vilaneuva, M. R., Smith, T.L., Erickson, J. S. (2003). Pain assessment for the dementing elderly (PADE): Reliability and validity of a new measure. *Journal of the American Medical Directors Association, 4,* 1–8. (http://blogs.lawyers.com/2013/09/antipsychotic-drugs-over-used-in-nursing-homes/).

Voyer, P., Martin, L. S. (2003) Improving geriatric mental health nursing care: Making a case for going beyond psychotropic medications. *International Journal of Mental Health Nursing, 12,* 11–21.

Warden, V., Hurley, A. C., Volicer, V. (2003). Development and psychometric evaluation of the pain assessment in advanced dementia (PAINAD) scale. *Journal of American Medical Directors' Association, 4,* 9–15. (Developed at the New England Geriatric Research Education & Clinical Center, Bedford VAMC, MA).

Weiner, D., Peterson, B., Ladd, K., et al. (1999). Pain in nursing home residents: an exploration of prevalence, staff perspectives, and practical aspects of measurement. *Clinical Journal of Pain, 15,* 92–101.

Wong-Baker FACES® Pain Rating Scale used with permission from the Licensing Department Wong-Baker FACES® Foundation. http://wongbakerfaces.org/

Chapter 4
Alzheimer's: Strips Individuals of All Skills

The tragedy of Alzheimer's is that not only is it a fatal disease but also it gradually and completely destroys all skills that we associate with adulthood and independence. This stripping away of skills includes the most basic including our ability to manage the minute details of our lives, our finances, our ability to shop and cook for ourselves, bathe, dress and toilet ourselves. No one would choose this.

Instrumental Activities of Daily Living

In a 2011 study that investigated the link between neurocognitive measures and instrumental activities of daily living (IADLs) in women and men with mild Alzheimer's disease (AD), the results of the study demonstrated that memory and executive functioning were related to IADL scores. Executive functioning was linked to total activities of daily living (ADL) abilities. Comparisons stratified on gender found that attention span predicted total ADL score in both men and women. Attention predicted bathing and eating ability in women only. Language ability predicted specific IADL functions in men (food preparation) and women (driving). The researchers concluded that even in patients with mild Alzheimer's, the ability to perform IADLs requires complex cognitive processes (Hall et al. 2011). These skills are lost in a relatively orderly and predictable sequence.

When an individual has deteriorated to stage 4 on the functional assessment staging tool (FAST) scale with a cognitive level of somewhere between 8 and 12 years of age, but before there has been a diagnosis of Alzheimer's, family and friends are noticing changes that are worrisome. Once a diagnosis is made, these same family members and friends recall the early signs that should have alerted them to a problem, but as the changes multiply with increasingly more serious consequences, denial no longer works and Alzheimer's becomes a reality. Looking back, family members can usually recall the individual losses that gave them cause for worry but somehow they ignored. Perhaps it was the inability to pay bills and manage the

V. Benner Carson et al., *Care Giving for Alzheimer's Disease*,
DOI 10.1007/978-1-4939-2407-3_4

checking account that first gave them reason to pause. Or, perhaps their anxiety stemmed from their Mother's inability to cook anything more than a cup of soup and a grilled cheese sandwich, when they had always known her to be a "world class" cook. Or, perhaps the worry arose when they saw a laboratory report that indicated Mom's blood sugar was very high, when for years she had done a wonderful job of managing her diabetes. However, as these losses begin to multiply along with the resultant problems from the inability to manage the checkbook; the inability to shop and cook meals that are "diabetic appropriate"; the getting lost in a neighborhood where the parent not only grew up but also raised a large family—denial fails! Loving family members will insist that their loved one see a physician. In fact, some family members will either insist on accompanying that person to the physician's office or will contact the physician prior to their loved one's appointment to fill the physician in on the problems that the family is seeing. What the family is observing is the slow degradation of many of the activities and skills that accompany "being an adult." Alzheimer's is responsible for these losses and more.

First, the IADLs, including food preparation, housekeeping and laundry, managing financial matters, shopping, using the telephone, taking medication, and handling transportation—all of these skills deteriorate then disappear. Before most individuals have even received a diagnosis of Alzheimer's, family members and close friends have noted changes. A daughter is puzzled when she calls her mother to get her mother's cheese cake recipe, only to be given vague directions. A son stops by just to say hello to his father and becomes concerned about the large pile of unopened and presumably unpaid bills piled on the kitchen table. Almost imperceptibly, the skills of being an independent adult disappear.

IADL and ADL

The IADLs include the following skills: managing finances, handling transportation—either driving or navigating public transportation, shopping, preparing meals, using the telephone and other communication devices, managing medications, and performing housework and basic home maintenance. The ADLs include bathing, dressing, and toileting. Let us examine what happens to each of these skills in stages 4, 5, and 6 on the FAST scale.

Stage 4: Mild AD: cognitive level 8–12 years of age.

The diagnosis of AD can be made with considerable accuracy in this stage. The most common functioning deficit in these patients is a decreased ability to manage instrumental (complex) activities of daily life. Examples of common deficits include decreased ability to manage finances, to prepare meals for guests, to take medications as prescribed, to schedule and follow-up for medical and other appointments, and to go to the market for oneself and one's family. The stage-4 patient has difficulty writing the correct date and the correct amount on the check. Consequently, someone either needs to supervise or assume this responsibility. For the stage 4 individual who lives independently, this difficulty may become apparent in responsibilities such as paying the rent and other bills. A spouse may note difficulties in writing the correct date and the correct amount when the affected spouse is writing checks. The stage-4 person is at greater risk for being taken advantage of by unsavory individuals who prey on individuals with Alzheimer's and use dishonest means to gain access to the individual's property, money, and other/or valuables. Writing a grocery list, as well as independently going to the market to buy food and other products is compromised in this stage. Preparation of meals, even for the best cook, becomes compromised. The individual begins to shy away from entertaining others because of a growing awareness of her/his deficits in this area. Even the ability to order food from a menu in a restaurant begins to be compromised. It is not uncommon for the person with AD to hand the menu to a family member or friend and instruct that person to order the meal. During this stage, the individual is still able to dress self independently and with good quality. Although he/she is having difficulty managing the IADLs, ADLs are still appropriate.

Stage 5: Moderate AD—cognitive level on FAST scale 5–7 years of age.

During this stage, when the person is functioning at a cognitive level between 5 and 7 years of age, deficits become even more marked. At this stage, deficits are of sufficient magnitude to prevent independent, catastrophe-free, community survival. Someone needs to shop for and prepare the food, ensure that the rent and utilities are paid, and assist in managing medications—not only ensuring that medications are taken as prescribed but also ensuring that prescriptions are filled. At this level, those with Alzheimer's can still dress themselves but not always with good quality—the clothing may be poorly matched, unwashed, and inappropriate for the weather or the occasion.

When basic ADLs start to be affected, however, this can present a real challenge for the caregiver. Let us take a look at a case study that presents the challenges that are inherent in bathing someone with AD.

Case study: as told to us by Joan, a family caregiver in California.

My Grandmother moved in with our family when she began to wander and we were concerned for her safety. Her son is my Dad. Besides my Grandmother, there were three of us in the house—me, my Mom, and My Dad. My Grandmother was always a very loving, spirited person. She paid immaculate attention to her grooming. After she had Alzheimer's for a few years, she cared less about grooming and was agitated around bathing when she

moved in with us. When we tried to bathe her she either cried or started screaming. My parents worked long hours, so as the granddaughter I was charged with her personal care. Every time I said, "Come on Gram, let's take a bath," she got agitated and refused. Showering was absolutely out of the question for her. I eventually realized that in order to bathe her, I was going to have to make this a special event for her. All her life she loved Motown music and going to her local hairdresser which she thought of as a spa. An idea dawned on me one day—why not make the bathroom like a spa with her favorite music?
So I got the bathroom ready—put a nice bubble bath in the tub, lit candles, and had her favorite music playing. As I asked Gram if she would join me for a Spa Day, she replied "Yes" and I led her to the bathroom. I bathed her and sang to her. The bathroom was warm and she seemed to be enjoying herself. I was soaked when we were done bathing but she was clean and there was no resistance to bathing. After being frustrated for weeks trying to bathe Gram, I finally realized I had to change how I approached and bathed her as well. She began to look forward to the "Spa Days" with me. I look back and realize it just took me adjusting my approach to bathing Gram.

Stage 6: moderately severe AD—cognitive level on the FAST scale—4–2 years of age.

Throughout this stage, ADLs continue to deteriorate. Individuals in stage 6 initially need assistance with bathing, dressing, care of their teeth, and eventually toileting. By the end of stage 6, all of the ADLs are completed with maximum assistance. The abovementioned case study illustrates a creative approach to bathing.

One of the "rules of thumb" when helping a loved one with their basic ADLs is to establish and maintain a routine—activities should be done at the same time and in the same order each day. "Sameness" and routine seem to help those with AD to remain calm.

Bathing

Let us first examine bathing—it seems to be one of the most challenging tasks for caregivers and those with AD. Why is bathing a battle ground? What makes the person with AD so resistant to an activity that he/she freely and independently engaged in prior to developing Alzheimer's? Is it fear, pain, embarrassment, or some other issue that leads to aggressive behavior? Indeed, it may be all of these explanations. When someone requires total assistance to perform basic ADLs (bathing, dressing, brushing of teeth, eating, taking medications, toileting), the person has deteriorated to stage 6 on the FAST scale with a cognitive and functional level of a toddler—somewhere between 2 and 4 years of age. The question then is, "Do we generally shower toddlers?" Unless the parent is holding the child in the shower while they are both being showered, the answer is a resounding no! Showering is frightening to a small child—the water hitting that child in the face feels scary. This is also true for an adult who is in stage 6 on the FAST scale. And yet, showering is the preferred method of bathing for both family and paid caregivers. Why do caregivers persist with an activity that results in anxiety for the recipients of their care as well as places them in a dangerous situation? Probably, the sense of being overwhelmed blocks most creative thought on the part of caregivers (Carson 2011a, b).

On the surface, showering seems to be the easiest and most efficient way to bathe another adult. Indeed, this adult probably showered him/herself when that was possible prior to the cognitive and functional decline of Alzheimer's. As a result of this decline, the person with Alzheimer's gradually loses all the abilities that normally accompany being an independent adult and increasingly demonstrates behaviors and responses characteristic of young children. If Dr. Seuss, who wrote so many wonderful children's books, were to write about this topic, the verse might be "I can bathe you here, I can bathe you there, and I can bathe you just about anywhere—except the shower." (See Appendix A for list of bathing resources.)

Let us explore the alternatives not just for bathing, but for other ADLs. First, let us look at bathing. If the person requiring care does not have arthritis in multiple joints and is able to get in and out of a tub, this may be the best way to bathe him/her. However, if arthritis is present, medicating the person with Tylenol© or an anti-inflammatory drug 30 min prior to the bath might allow sufficient pain relief that the person is able to independently get into and out of the tub. If over-the-counter drugs do not provide sufficient pain relief, then a prescription analgesic is probably in order. Otherwise, there are quite a few alternatives to a tub bath, as we describe below.

Preparation is the key to successful bathing. Towels that will be used as part of the bathing experience should be warmed in the clothes dryer prior to bathing so that the person being bathed can be wrapped in warmth throughout the bathing experience. Before escorting the person with AD into the bathroom, it helps to fill the bathroom with a "head of steam." The damage to the hypothalamus that controls body temperature leads to intense sensitivity to cold, and once someone with AD becomes chilled it takes a while for that person to feel comfortable again. Mirrors should be removed or covered because it is frequently frightening for the person with AD to see his/her own reflection in the mirror. That reflection may not be recognized by the person with AD and might be interpreted as "a stranger is in the bathroom with me!"

Bathing can be accomplished standing at a sink, wrapped in warm towels for comfort as well as modesty. Bathing can be accomplished with the person sitting on the toilet which allows almost complete access to all body parts. Bathing can be done in bed or in a chair. The caregiver needs to consider the bathing products that are used. For example, Ivory® soap might have been the only option available when the person with Alzheimer's was a young person, thus the use of Ivory® soap might make the bathing experience a bit more pleasant. It is important that the caregiver become as knowledgeable as possible regarding the "habits" that might have been part of the early history of someone who has AD. For instance, if the person receiving care only bathed on Saturday in preparation for church the next day when he/she was a young person, then the caregiver should make every day "Saturday" when bathing is needed! Some of the person's deficits due to Alzheimer's can be resources to the caregiver faced with challenging behaviors.

Still another strategy for bathing is to use a distractor such as candy. People in stage 6 can only do one thing at a time—so if sucking on a Tootsie Roll Pop® or other candy serves as an effective distractor, then that might be the strategy to use. Candy is certainly easy to obtain and allows the caregiver to "get the job done." Still another strategy is to use music. Music has many benefits. It can be employed as a distractor when there is a necessary but unpopular activity that must be accomplished i.e., bathing. Music is useful to reengage the person with AD in life and also as a way to sooth and comfort. Physicians should write a prescription for the use of music multiple times a day, every day. This one intervention could be very effective in reducing challenging behaviors. The case study presented at the beginning of this chapter illustrates the power of music in the bathing experience.

Dental Care

Maintaining dental health is another challenge for caregivers. Regular tooth brushing needs to be supervised by the caregiver. Sometimes using a "watch me" technique is useful. The caregiver demonstrates step by step how to brush teeth and encourages the person with AD to follow suit. Sometimes it is necessary for the caregiver to place her/his hand over the hand of the person with AD and gently guide the toothbrush in and out of that person's mouth. It is essential that the caregiver monitor daily oral care. Encouraging and/or participating in making sure that teeth or dentures are brushed every day along with flossing is essential to maintaining dental health. Brushing teeth or dentures after each meal is important. If the caregiver uses disposable flossing devices, the task of flossing becomes easier. Dentures need to be removed and cleaned every night followed with gentle brushing of the person's gums, tongue, and the roof of the person's mouth. It is important for the caregiver to investigate any signs of mouth discomfort that appear during mealtime. The person with AD may refuse to eat or make strained facial expressions while eating. These signs may point to mouth pain or dentures that do not fit properly. As the disease of Alzheimer's progresses, self-care skills and interest in most aspects of self-care tend to diminish. The person may lose the ability to brush his/her teeth. The caregiver needs to seek advice from the dentist as to the best way to clean the person's teeth. Many caregivers find that it is easiest to sit the person in a chair and to brush their teeth from behind. In this way, the head can be rested back on the chair, or the lap of the caregiver, and the head moved with relative ease.

It is important to keep in mind that the person's ability to describe dental symptoms or pain diminishes as dementia progresses. Apart from maintaining healthy teeth, it is important to check the mouth for damage to teeth, gums, or the tongue (biting for example). The chance of mouth cancer also increases with age, but if caught early responds well to treatment. Another area that demands the attention of the caregiver is awareness of behaviors that might indicate dental problems. The person with Alzheimer's will not be able to report a toothache. Behaviors such as rubbing or touching the cheek or jaw, moaning or shouting out, head rolling or "nodding," flinching—e.g., when washing the face or being shaved, refusing hot or cold food or drinks, restlessness, poor sleep, increased irritation or aggression or refusal, and/or reluctance to put in dentures when previously this was not a problem are all behaviors that indicate the need for a dental evaluation.

Dressing

The person with AD may not remember how to dress or may be overwhelmed with the choices or the task itself. The caregiver can provide assistance in a number of ways. First, by simplifying the choices, for example, the caregiver chooses two pair of sweatpants and asks, "Would you like to wear the red ones or the blue ones?" The process of dressing needs to be organized for the person with AD. For instance, clothes need to be laid out in the order that they are put on the body. Without this organization, the woman with Alzheimer's might put her bra on over her sweater and a gentleman might wear his underpants over his sweatpants. Clothing should be comfortable and easy to put on and take off. Buttons, snaps, and zippers can be replaced with Velcro®.

The caregiver needs to simplify but make specific the available choices. Rather than asking, "What would you like to wear today?" which is too open-ended, the caregiver asks, "Would you like to wear the red or the blue sweat pants today?" If the person wants to wear the same outfit day after day, it is helpful for the caregiver to purchase multiples of that outfit so that the person feels in control while the caregiver is able to wash the clothes. Purchasing comfortable shoes with nonslip soles is also an important consideration, especially as balance becomes a problem. Finally, because ADL can prove to be very frustrating for the person with Alzheimer's, it is important for the caregiver to be both patient and prepared to simplify the process.

The following case study was shared by a home care provider:

> Joe, a home care nurse, received a referral for a gentleman with Alzheimer's who was demonstrating challenging behaviors. As Joe got ready to knock on the door of this patient's home, Joe heard yelling coming from inside. He waited until the yelling subsided before he knocked. The door was opened by the patient's wife who appeared frazzled. After Joe introduced himself as the home care nurse, the wife began to describe her frustrations. She explained that every single aspect of care was difficult because her husband "refuses to help me—even in the tiniest way". She proceeded to give an example. "If you heard me yelling before you knocked it was because he refused to help me put his shirt on. I helped him put his arm into one of the sleeves and then I said put your other arm in the other sleeve—easy huh? Three times in a row he struggled with the shirt and then finally removed it from the first arm where I had helped him. I don't know what to do—this is so frustrating and it is sure not the retirement that I envisioned when we moved from Chicago to Phoenix. The doctor said you could help me. I sure hope so because I am at my wits end." Joe listened patiently and explained that perhaps her husband didn't have enough short term memory to remember how to follow her directions and maybe an easier solution would be to get rid of shirts with buttons and use pull-over shirts instead. Joe spent that first visit gathering information from Nick's wife and was able to assess that the patient was in the beginning of Stage 6 on the FAST scale. Nurse Joe explained what that meant in terms of Nick's capabili-

ties, limitations, and strategies to be used by his wife. Before that first visit was over, the nurse noticed that there was an organ in the living room. He asked the wife "who plays the organ?" The wife replied that Nick, who used to be active in their church, was the organist, but she assured the nurse that her Nick could no longer play the organ. Joe questioned that conclusion and walked the patient into the living room with the organ. The patient sat down at the organ and immediately started to play "Amazing Grace". The wife began to cry. She was astonished that her husband still retained this ability. Joe the nurse assured her that many of the "old skills" might have been "overlearned" and were still available to the person with Alzheimer's. Joe encouraged the wife to have her husband continue to play the organ on a regular basis because it would be a source of joy for both of them. Over the course of two months, Joe taught the wife techniques to manage her husband's care. She simplified bathing by showering with her husband. She changed his clothes from button shirts to pull-overs and sweat pants that took the place of dress pants that required a zipper and a belt! About six months after Joe had discharged the patient, he received a telephone call from the gentleman's wife. She wanted Joe to know that her husband had recently died—he fell in the driveway and developed a sub-dural hematoma that was followed with many complications. She told Joe that his instruction had allowed her to have six months of peace with her husband—the frustration and yelling had disappeared, and she discovered that her husband still retained the ability to laugh easily and to play the organ, both of which had given them great pleasure. She expressed gratitude for the blessing of those six months.

Toileting

The ability to independently use the bathroom is lost at the end of stage 6. At stage 6c, the person loses the ability to handle the mechanics of toileting which means that he/she no longer can wipe her/himself, flush the toilet, or properly dispose of the toilet tissue. At stage 6d, the person becomes incontinent of urine and at 6e the person is incontinent of stool. At these levels on the FAST scale, the cognitive ability is that of a toddler about 2 years of age.

Frequently, when caregivers recognize that the cared for person is "making a mess" in the bathroom because he/she is not sitting on the toilet or standing properly in front of the toilet, or that the person is using a whole roll of toilet paper with more toilet paper on the floor of the bathroom than in the toilet, the response of the caregiver is to purchase some type of incontinence product. The problem with the premature use of these products is that the caregiver may leave a soaked adult "diaper" on too long, increasing the risk of a urinary tract infection. When the person with Alzheimer's begins to demonstrate difficulties with the mechanics of toileting, the caregiver needs to establish an every 2-h toileting schedule in much the same way that a parent would teach a toddler to use the bathroom. By initiating a trip to the bathroom or encouraging the person with Alzheimer's to use the bathroom every 2 h during the day, beginning with a trip to the bathroom on awakening, the caregiver can cut down on the cost of incontinence products and allow the person to remain just a bit more independent. Another consideration that must accompany a timed voiding schedule is that the person with Alzheimer's needs to receive a drink of juice, water, tea, or coffee every 2 h. This strategy will be quite effective in eliminating the risk of the person developing a urinary tract infection which could lead to delirium and even death. The reality is that no matter what strategies the caregiver employs the cared for person will become incontinent. The goal is to manage that incontinence in the best way possible, moving to a timed voiding schedule, and incorporating regularly scheduled intake of fluid.

Walking

By the end of stage 6, the person will be struggling to walk and may have even experienced a hip fracture as a result of a fall. Again, the individual is functioning at the level of a toddler at this point. Toddlers can walk but their balance is usually not good and they frequently fall. The difference between the toddler falling and the elderly person with AD falling is that the elder has a greater distance to fall and her/his bones are brittle so that falls carry with them the risk of hip or other bone fracture. If the elderly person sustains a hip fracture without additional complications, the hip can be surgically repaired. The problem arises with the use of a walker. People in stage 6 lack the capacity to learn anything new—they only have a few minutes of short-term memory—so it is not likely that this person will consistently, if at all, use the walker. Without the walker, the person will fall again. For many with Alzheimer's, a fractured hip signals the beginning of the end.

What is the answer? The answer is to provide a walker that has a cart on the front and resembles the shopping care that the elderly person has pushed through stores for many years—this is drawing on "old" memories. The "cart" rather than the "walker" has a sign on it that instructs the person—"don't forget to use your cart." Every room should have a reminder to the person with the healing hip that he/she needs to use his/her cart!

Summary

This chapter presents details regarding the significant losses that occur due to AD. No one would argue with the statement that Alzheimer's is a devastating disease—both for the person afflicted with the disease, that person's spouse and family and in fact for everyone who knows the individual. The losses are slow, they are irreversible and continue until death, they strip away everything that we associate with adulthood and independence, and although the person deteriorates to the level of an infant, there is none of the joy that accompanies a baby. The disease provides a frightening glimpse into a possible future for all of us—a very unwelcome future.

Appendix A

Products for hair washing
www.carepathways.com go to Shop Medical
(see No Rinse Shampoo, Hair Washing Tray, EZ-Shampoo Shower, EZ-Shampoo Basin)
www.allegromedical.com
(see Hair Washing Tray, Hair Wash and Rinse, Tray with Spray Nozzle)
DVD/CD on bathing
"Bathing without a Battle" http://bathingwithoutabattle.unc.edu/
"Bathing without a Battle" book on amazon.com

References

Carson, V. B. (2011a). Responding to challenging behaviors in those with Alzheimer's: Communication matters. *Caring Magazine, 30*(3), 26–31.
Carson, V. B. (2011b). Taking the angst out of bathing. *Caring Magazine, 30*(4), 40, 42.
Hall, J. R., Vo, H. T., Leigh, A., Johnson, L., Barber, R. C., O'Bryant, S. E. (2011). The link between cognitive measures and ADLs and IADL functioning in mild Alzheimer's: What has gender got to do with it? *International Journal of Alzheimer's Disease, 2011,* 276734. doi:10.4061/2011/276734. (6 pages, 2011).

Chapter 5
Help! My Mother Is Lost and I Cannot Find Her!

Wandering is a common behavior that occurs in Alzheimer's disease (AD)—the Alzheimer's Association estimates that over half of all those with the disease will wander at one time or another and if the wanderer is not found within 24 h, up to half of individuals who wander will suffer serious injury or death (Alzheimer's Association 2014). Wandering results from brain damage in the areas of the parietal lobe as well as the hippocampus (see Chap. 2). Wandering involves the need to "keep on the move" (Alzheimer's Association of Canada 2013).

Regardless of whether or not the caregiver is paid or is a family member, the behavior of wandering "strikes fear" into their hearts. In order for caregivers to experience a sense of competence in the caregiving role, they must be able to institute strategies that keep the cared-for person safe. This chapter provides relatively simple and inexpensive tips for caregivers to limit wandering. For example, an intervention such as placing a black or brown rug held in place with electrical tape in front of all exits of a home is very effective for some with AD. The person with AD loses depth perception and for many the dark rugs will appear as "holes"—the person may be too afraid to cross over the "hole."

In Chap. 2 we introduced the idea of wandering with a case study of a Virginia man suffering from Alzheimer's who managed to get on a bus in Virginia and travel to Denver, Colorado. That story had a happy ending because two kind police officers took the time to make sure the gentleman was cared for while they made arrangements to send him safely home. He was a very fortunate man. Not all of the stories about wandering, though, have a happy ending.

How many of those with Alzheimer's will wander? The statistics are imprecise but the numbers reported by different chapters of the Alzheimer's Association are about five to six of every ten people with the disease. There are a number of facts that make wandering of even greater concern. A person with Alzheimer's may not remember his or her address, easily become disoriented in familiar as well as unfamiliar locations, and may not respond to searchers calling his/her name. In fact, the person could be standing in a dark doorway, just feet away from a searcher and fail

V. Benner Carson et al., *Care Giving for Alzheimer's Disease*,
DOI 10.1007/978-1-4939-2407-3_5

to identify him/herself to the searcher. One question might be how this behavior fits into the theory of retrogenesis. A story from a mother of a 3-year-old might shed light on this issue:

> I was shopping in a K-Mart store and my three year old son Robbie was right alongside of me. I stopped to look closely at some items and I let go of Robbie's hand. I am not sure how much time elapsed before I realized I was not holding his hand but I believe it was only two or three minutes. When this realization hit me, I immediately called his name; I looked all around and I didn't see him. I panicked! I had the store manager lock down the store and he called Robbie's name over the PA system. I think it might only have been about two to three minutes but it seemed like an eternity to me before Robbie came out of the center of a skirt rack and asked whether I was looking for him? I had looked under the skirts but in the center of the rack there was a raised area, just big enough for Robbie to balance on and covered by skirts so I did not see him when I had looked. When he asked if I was looking for him, I had mixed feelings—relief and joy that he was okay but also anger that he had caused me such anguish. He loved to hide from me in this way—and tried to do it multiple times, but the K-mart incident had taught me a valuable lesson.

What does Robbie's story have to do with the wandering of an older adult with Alzheimer's? How does this fit in with the theory of retrogenesis? The response or lack of response of the older adult with Alzheimer's to hearing his/her name called is frequently the same as Robbie's toddler response—they do not call out, "here I am" or otherwise make themselves known—which is why wandering is such a dangerous behavior. To make matters worse, the person with Alzheimer's may not remember vital information like his/her name or address and may become disoriented even in familiar places.

Who Is at Risk?

Anyone who is diagnosed with AD and can walk is at risk for wandering. Certainly the risk increases as the dementia progresses but even at stage 4 on the functional assessment staging tool (FAST) scale, people may still be driving (albeit with questionable skill) and can easily become confused about directions. In fact a wake-up call for many families who have noticed a loved one's repetitive questions and stories occurs when the loved one "gets lost" attempting to drive home from a familiar location. The following is a real life scenario:

> Mrs. Johnson goes to the local grocery store where she has shopped for years. However, she becomes confused when leaving the parking lot, turns right instead of left and drives for hours. Her home is only a five minute drive from the grocery store she just left. Finally, she is pulled over by a police officer as she is weaving across the double lines of the highway—80 miles from her home. Fortunately, she has an emergency contact card in her purse with her daughter's information. The police officer calls the daughter to inform her that her mother is quite disoriented and that they are taking her to the closest hospital for an evaluation. The daughter can no longer tell herself that her mother "has a little memory problem" and the daughter will now be faced with a myriad of decisions regarding what is best for her mother. How can her mother be kept safe? What supports does mom need and who will provide those supports? And of course, must confront her mother and ask her to relinquish the keys to the car.

Sometimes stories like Mrs. Johnson's often make the national news. Even in the early stages of dementia, a person can become disoriented or confused for a period of time. It is important to plan ahead for this type of situation. The following are warning signs that the loved one is having difficulty getting around safely:

1. Returns from a regular walk or drive later than usual
2. Tries to fulfill former obligations, such as going to work
3. Tries or wants to "go home," even when at home
4. Is restless, paces, or makes repetitive movements
5. Has difficulty locating familiar places like the bathroom, bedroom, or dining room
6. Asks the whereabouts of current or past friends and family
7. Acts as if doing a hobby or chore, but nothing gets done (e.g., moves around pots and dirt without actually planting anything)
8. Appears lost in a new or changed environment (Alzheimer's Association 2007)

Suggestions to Prevent Wandering

Wandering can happen even if caregivers are extremely diligent. It is important for the primary caregiver to establish a daily routine of activities to provide structure to the day, decrease agitation and improve the mood of the person with AD. It is important when planning activities for a person with dementia that the caregiver considers the likes and dislikes, the strengths, abilities, and interests of the individual and is willing to continually explore, experiment, and adjust these activities. How did the person used to structure his/her day? At what time of the day is the person at his/her best? These, and allowing unrushed time for performing activities of daily living and for meals, are important considerations. Providing structure with flexibility is the best approach.

As AD progresses, the abilities of a person with dementia will change. With creativity, flexibility and problem solving, the caregiver will need to adapt the daily routine in response to these changes. The following is a checklist of daily activities that could allow the caregiver to structure the day for a person with AD.

Checklist of Daily Activities to Consider

Household chores—dusting, sweeping
Mealtimes—setting the table and removing the used dishes at the end of the meal
Personal care; bathing, dressing, brushing teeth, toileting
Creative activities (music, art, crafts)
Intellectual activities (reading, puzzles)

Physical activities—taking daily walks, raking leaves, assisting in the garden, putting silverware away, folding laundry and others

Social activities—visiting with family members and close friends, shopping with a responsible person,

Spiritual activities—attending choir practice, attending religious services, reading Scripture, listening to hymns and religious services on the television and/or radio, praying

It is beneficial for the primary caregiver to develop a daily plan—this does not rule out spontaneity in the day's routines, but a solid plan gives the caregiver structure that she/he can rely on when challenges arise that make it difficult to think creatively. A plan can always be modified because of a change in circumstances. It is much more difficult to "come up" with a plan when the caregiver is feeling stressed because the loved one with AD is displaying a new challenging behavior.

Writing the Plan

Another useful strategy is that the daily plan be written and posted in a highly visible place so that the caregiver and the person with AD can refer to the plan throughout the day. Included in the plan are many activities that those of us who do not have a brain altering illness need no reminders to complete. For instance, the morning plan could look something like this:

- Wash, brush teeth, get dressed
- Prepare and eat breakfast
- Have coffee, make conversation
- Discuss the newspaper, try a craft project, reminisce about old photos
- Listen to calming music
- Take a break, have some quiet time—could be a time for quiet prayer or reading of religious material, i.e., Bible, or looking at old picture albums that frequently serve as an impetus to conversations about the person's younger years. These conversations provide the caregiver insight into the person with Alzheimer's history.
- Do some chores together—dusting, sweeping, folding laundry—any mindless, repetitive activity will do
- Take a walk outside if weather permits—otherwise walk in the house, play an active game—men love to manipulate and build with Legos© or sort coins; women may enjoy winding a ball of yarn; coloring, drawing; both men and women enjoy listening to music from his/her era and/or watching old movies.
- Prepare and eat lunch, read mail, wash dishes
- Listen to music, do crossword puzzles, watch TV
- Clip coupons
- Do some gardening, take a walk, visit a friend

- Take a short break or nap (http://www.alz.org/care/dementia-creating-a-plan.asp#organizing#ixzz301pelEey. Accessed 26 April 2014).

Reassurance provided regularly is essential to the person with AD because this person might frequently feel lost, abandoned or disoriented. Many of those with this terrible disease will insist that he/she wants to go home or to go to work. The caregiver needs to explore and validate. For instance, in response to a father's arguing that he will be late for work, the daughter can validate her father's need, "Dad you must really miss work—you were always such a terrific worker. Tell me what your favorite responsibility was when you went to work."

Also important is avoiding the need to correct the person. Sometimes this just generates anger from the person with AD. It is better to provide a positive statement about what is going to happen. For example, "We are staying here tonight. We are safe and I'll be with you. We can go home in the morning after a good night's rest."

Additional Interventions

The following suggestions are focused on the perspective that the person who wants to "get out" might be experiencing unmet physical or emotional needs. It is important for the caregiver to ensure that basic needs like going to the bathroom, drinking, and eating are met. Those with AD will wander in search of getting these needs met. It is important to avoid taking the person with AD to shopping malls and other busy places; such situations lead to confusion and disorientation.

Camouflaging doors and doorknobs is another useful strategy to decrease wandering. This can be accomplished by painting the doors the same color as the walls or by covering the doors with removable curtains or screens. Another method of camouflaging exterior doors is to use wallpaper that appears to be a library. The person with AD begins to lose depth perception and can no longer distinguish the door. Some assisted living facilities (ALF)'s wallpaper the area that surrounds the exit doors so that the door is disguised and appears to be a bus stop. Frequently the facility will add a bench alongside of the door and residents will come and sit on the bench and tell others that they are waiting for the bus to arrive (Kincaid and Peacock 2003). These strategies are using the deficits that accompany Alzheimer's in a creative fashion that keeps the person with AD safe. Additional strategies are to use black or brown rugs in front of every exit—the person will interpret these rugs to be dark holes and will not walk over them to exit. Still another safeguard is to use signaling devices such as a bell placed over a door or an electronic home alarm that signals when anyone tries to exit the door. It is important that a person with AD is never left alone in a locked home or in a car without supervision. Car keys must be kept safely stored in a place that is inaccessible to the person with AD; otherwise, the person might drive off and be at risk of potential harm to self and/or others.

Night Wandering

If the person with AD is a night wanderer, there are a few strategies that may help to keep that person safe. First of all, it is important to restrict fluids 2 h before bedtime and ensure that the person has gone to the bathroom just before bed. The use of night lights throughout the home helps to keep the night wanderer safe. It is also important to make sure that the bathroom is not only easily accessible but also that the colors in the bathroom support the person's ability to distinguish the toilet from the other items that are in most bathrooms. The following case study will make this last point clear.

> A family sought advice about their mother who had Alzheimer's. Every night she would awaken in the early morning hours and make her way to the bathroom. When she got to the bathroom she urinated in the trashcan. Why? Because everything in the bathroom was white except for the blue trash can! This problem was easily remedied with a bright blue toilet seat.

In addition, the locks on all doors leading to the exterior of the home need to be moved to the top or the bottom of the door—the person with AD will only look for the lock in the middle of the door. The door leading to the basement also needs to be locked. Medications as well as cleaning supplies and any other substances that the person with AD could consume during his/her nighttime wanderings need to be safely stored and locked up). It is helpful for the primary caregiver to walk through the house with an "eye" to noticing anything that would pose a risk to a nighttime wanderer. As the caregiver makes the rounds of the house it is helpful for him or her to keep in mind this idea, "If I had a toddler, what would I want to remove so that my child did not come in contact with any substance that could cause harm?" The person with AD may be functioning at the level of a toddler (stage 6 on the FAST scale) so the same precautions need to be in place.

Sometimes a simple intervention like posting "Stop" signs or "Do not enter" signs on doors that lead outside may be enough to keep a loved one from wandering.

Developing a Neighborhood Watch

Introducing a loved one with AD to the neighbors is a good idea. The neighbors need to be familiar with the person with AD so that if they see this person wandering they know that this is an unsafe behavior. Following the introduction, it is a good idea to send a "Dear Neighbor" letter to reinforce that idea that the loved one has a memory problem and should not be out alone. This is an easy and effective way to set up a "neighborhood watch." If the person with AD is able to get out into the community without the supervision of the family caregiver(s), neighbors are asked to invite the "wanderer" in for a chat and to inform the family of the loved one's location. The primary caregiver needs to have the telephone numbers of people that can be called upon to assist in a search for a loved one who is missing. It is important that search and rescue efforts begin as soon as possible. Once a caregiver and others have searched the immediate area for 15 min, 911 needs to be called. Research shows that 94 % of people who wander are found within 1.5 miles of where they disappeared (DBS Productions 2012).

Creating an Emergency Plan

Sometimes, despite the most careful and observant caregivers, the person with AD gets out of his/her home or the ALF and wanders. The stress experienced by families and paid caregivers when a person with dementia wanders and becomes lost is significant. It is essential to have a plan in place beforehand, so that if the person wanders a plan is immediately activated. The following suggestions are useful when faced with the reality that a loved one with AD has wandered away and the caregiver does not know where. It is vitally important that caregivers are prepared for possible wandering in the following ways:

- Have available a recent, close-up photo and updated medical information for police.
- Ensure the person with AD always carries/wears identification—medical ID jewelry—like a bracelet or a pendant.
- Be knowledgeable about the neighborhood—specifically areas that pose a danger such as bodies of water, open stairwells, tunnels, bus stops, dense foliage, roads with heavy traffic;
- Be aware of whether the individual is right- or left-handed. Wandering generally follows the direction of the dominant hand.
- Be aware of places that might hold significance and where the wanderer might try to access such as past jobs, former homes, places of worship, or a restaurant.

The individual needs to be enrolled in the MedicAlert® + Alzheimer's Association Safe Return® program (see Chap. 2). Additionally there are products on the market that use an electronic tracking global positioning system (GPS) device to allow the caregiver to locate the person.

Wanderers are often looking for something or someone familiar, especially if they recently moved to a new environment. In other cases, wanderers are trying to satisfy a basic need, such as hunger or thirst—but they have forgotten what to do or where to go. Many wanderers are looking for a bathroom. Some are trying to escape from something—stress, anxiety, or too much stimulation. For others the wandering may be linked to a lifelong routine, for instance, a woman who tries to leave every day at 6 a.m. may be trying to get to a job that she held for over 40 years. Caregivers may never know for sure why someone wanders but regardless of the reason for the behavior, it is a risky behavior and caregivers must think through strategies to deal with it.

Summary

Wandering is a common and very worrisome behavior that accompanies AD and other dementias. There is no clear-cut reason behind wandering. One common theme is that the person with dementia is trying to get someplace else and many times that "someplace else" represents a memory from his/her youth. The sought-out place may no longer exist except in the mind of the "wanderer." There are quite a few strategies to minimize the risk of wandering but even when all of these strategies are employed, some individuals with dementia find a way to get outside of their residence, seeking what is only known to that individual, and perhaps putting him/herself in danger. One of the important interventions is registering a loved one in the *Safe Return Program* sponsored by the National Alzheimer's Association.

References

Alzheimer's Association. (2007). Wandering behavior: Preparing for and preventing it. http://www.alz.org/living_with_alzheimers_wandering_behaviors.asp. Accessed 30 April 2014.

Alzheimer's Association of Canada. (2013). www.alzeimerbc.org. Accessed 26 April 2014.

Alzheimers Association. (2014). http://www.alz.org/care/alzheimers-dementia-wandering.asp#ixzz300tXKhrz. Accessed 26 April 2014.

Cloutier, D. (2013). Exploring the influence of environment on the spatial behavior of older adults in a purpose-built acute care dementia unit. *American Journal of Alzheimer's Disease Other Dementias*. doi:1533317513517033. Accessed 4 May 2014. (Thomson Reuters; first published on 31 Dec 2013)

DBS Productions (2012). The source of search and rescue research, publications and training. http:/www.dbssar.com/SAR_Research?Wandering_Characteristics.htm. Accessed 1 May 2014.

Kincaid, C., & Peacock, J. R. (2003). The effect of a wall mural on decreasing four types of door testing behaviors. *Journal of Applied Gerontology, 22*(1), 76–188.

O'Neill, D. (2013). Should patients with dementia who wander be electronically tagged? No. *British Medical Journal, 346*, f3606. doi: http://dx.doi.org/10.1136/bmj.f3606 (Published 20 June 2013). Accessed 2 May 2014. (Cite this as: BMJ 2013;346:f 3606 Desmond, professor of geriatric medicine).

Petonito, G., Muschert, G. W., Carr, D. C., Kinney, J. M., Robbins, E. J., Brown, J. S., & Muschert, G. W. (2013). Programs to locate missing and critically wandering elders: A critical review and a call for multiphasic evaluation. *The Gerontologist, 53*(1), 17–25. doi:10.1093/geront/gns060.

Chapter 6
Getting the Food in…and Getting it Out

Eating and elimination are part of the same continuum—and Alzheimer's disease (AD) impacts the entire continuum. In this chapter, we first address the issue of getting enough food into the person and how AD impacts appetite. This is a major concern for both family members and paid caregivers. They worry whether the person in their care is taking in enough calories and losing weight or is taking in too many calories and gaining weight, both of which may negatively impact their health. Perhaps the person under care is a diabetic or has any number of chronic diseases where dietary management is essential. Looking at where the person is on the functional assessment staging tool (FAST) scale, puts these issues into perspective.

With the onset of stage 6, when cognitive and functional ability is at the level of a 4-year-old deteriorating to that of 2-year-old, eating progresses through two major changes. At the beginning of stage 6, people are hungry all the time. They have very little short-term memory and quickly forget that they have just eaten. A caregiver might complain, "No sooner have I cleared away the dishes than my mother is asking me when we are going to eat! I become really frustrated with her when I need to tell her repeatedly, 'Mom you just ate two scrambled eggs, several slices of bacon, toast and two donuts. You can't possibly be hungry!' What am I supposed to do—cook for her all day long?" Cooking all day is not necessary but a strategy that works in this situation is to make healthy foods such as cheese, fruits, and crackers available all day long so that the person can snack whenever he/she wants. At the end of stage 6 on the FAST scale and moving into stage 7, the person frequently refuses to eat or eats very little.

The refusal to eat is a frightening behavior to caregivers. This seems unnatural and leads to weight loss and an assortment of other health problems including increased risk for infection, reduced muscle strength, fatigue, and increased falls (Increasing links on food and depression).

V. Benner Carson et al., *Care Giving for Alzheimer's Disease*,
DOI 10.1007/978-1-4939-2407-3_6

Dehydration

People with dementia are also at risk for becoming dehydrated if they are unable to communicate their need for fluid, recognize that they are thirsty, or if they just forget to drink. Insufficient fluid intake can lead to headaches, increased confusion, constipation, and urinary tract infections (UTIs). In the case of UTIs, most of those with Alzheimer's dementia demonstrate no symptoms of a UTI. However, an untreated UTI can lead to delirium—a life-threatening condition. Negative outcomes resulting from inadequate fluid intake can worsen the symptoms of dementia. Providing a drink every 2 h throughout the day is a strategy that helps to ensure fluid intake. This will maintain hydration and may prevent constipation, UTIs, and other physiological changes that can result in delirium (Trivedi 2014).

Weight Loss

Over the course of the disease, most people with Alzheimer's dementia generally lose weight. The loss of weight is probably due to a number of factors such as diminished appetite, depression, difficulty cooking, problems with communicating or recognizing hunger, poor coordination, tiring more easily, and problems with chewing and swallowing. Swallowing difficulties can be referred to a speech pathologist and loss of weight to the physician who can refer the patient and caregiver to a dietitian. Untreated depression, which is very common in those with dementia, may also lead to diminished appetite and weight loss.

Persons with dementia may have problems communicating hunger or dislike for the food they have been given. They may communicate their needs through behaviors such as refusing to eat or holding food in their mouth. Providing choices of foods, using prompts and pictures, sometimes help.

The presence of pain might also be a reason for refusal to eat. Untreated pain anywhere in the body can lead to a loss of appetite. However, pain due to problems with dentures, sore gums, or painful teeth often has a negative impact on eating. Oral hygiene and regular mouth checks are important.

Fatigue can also be a cause for not eating or giving up part way through a meal. Fatigue can lead to other difficulties such as problems with concentration or difficulties with coordination. It is important to be aware of this and to provide the necessary support for eating when the individual is most alert. Eating small portions more regularly is better than having set mealtimes and again, the strategy of placing healthy finger foods in a convenient location may encourage all-day snacking.

Other Causes for Diminished Appetite

Changes to medication or dosage can result in appetite changes. Likewise, lack of physical activity may directly impact appetite. Those who are sedentary may not feel hungry, whereas the person who is always on the move (pacing, wandering, etc.,) might seem hungry all the time and may need additional calories. Constipation is a common problem and can result in the person feeling bloated or nauseous, making them less likely to want to eat. Strategies to prevent constipation include encouraging activity (if sedentary), offering fiber-rich foods (vegetables and bran-containing cereals), and providing plenty of fluids (Alz site-eating tips 2013).

Strategies to Encourage a Healthy Appetite

There are many ways to stimulate appetite and interest in food and drink. First, knowing the person, his/her routines, preferences, likes and dislikes, and other needs will lead to more successful caregiving. Families should communicate this information to paid caregivers. The following are key to getting persons with dementia to eat:

- Presentation of food is important. Foods that smell and look good are appealing. Sometimes providing different tastes, colors and smells may stimulate a poor appetite.
- Seizing on opportunities to encourage eating as opposed to being locked into a rigid eating schedule is essential. For example, if the cared for person is awake for much of the night then night-time snacks may be a good idea.
- Providing food that the person likes in small portions is often helpful.
- Trying different types of food, e.g., milkshakes or smoothies (often fortified with protein) may be another way to increase the intake of calories and necessary nutrients.
- Alzheimer's may impact the way food tastes, so experimenting with stronger flavors or sweet foods may be helpful.
- Food that becomes cold will lose its appeal. If the person is a slow eater, it makes sense to serve only small portions to keep food warm, or to use the microwave to reheat food.
- If the person is having difficulties chewing or swallowing, serving naturally soft food such as scrambled eggs or stewed fruit might help—avoid pureed food since it is generally not appetizing.
- If pureed food is necessary, speak to a dietician or speech and language therapist to discover strategies to make sure that food is not only nutritious but continues to taste good.
- Involving the cared for person in the mealtime preparations—perhaps setting the table or stirring the food as it cooks—can often stimulate appetite.

- Providing positive encouragement and gentle reminders to eat may help, although impatient nagging is not.
- Maintaining a relaxed, friendly atmosphere with some soft music is often helpful.
- Meals should be a time not only for eating and drinking but also an opportunity for social stimulation.
- Providing dessert even if the person does not eat his/her entire meal is recommended—a motto for caregivers should be: "Who Says You Can't Eat Dessert First!"

Additional Challenges

People with dementia may struggle to recognize food and drink, leading to malnutrition and dehydration. This deficit can be explained by damage to the brain, unfamiliar food, or how food is presented. If the person with dementia has problems with vision, he/she may not be able to see the food. Explaining what the food is and ensuring that the person is wearing correct glasses are helpful interventions.

Additionally, those with Alzheimer's have limited concentration, which means they may have difficulties staying focused on a meal until it is finished. A solution to this is to serve smaller more frequent meals and/or to encourage healthy snacking. People with dementia may struggle to handle cutlery or pick up a glass. They may also have trouble getting food from the plate to their mouth. A person with dementia may not open her/his mouth as food approaches and may need to be reminded to do so. There may also be other medical conditions that affect the person's coordination, for example, Parkinson's disease. If he or she is struggling with a knife and fork, it helps if the caregiver chops the food so it can be eaten with a spoon. The following points are further suggestions based on the concerns above that might make meal times more productive:

- If the person appears to have difficulty using cutlery, the caregiver may need to prompt the person and/or place his/her hand over their hand in order to guide it to their mouth.
- Providing finger foods such as sandwiches, slices of fruit, vegetables, sausages, cheese and quiche make eating easier, especially when co-ordination becomes difficult.
- The person with AD should be allowed to eat wherever he/she feels comfortable.
- Chewing may be difficult—the person with AD may forget to chew or hold food in his or her mouth.
- As noted above, good oral hygiene is important. This includes evaluating for mouth pain that might interfere with eating.
- Eating soft, moist foods that need minimal chewing can help.
- Occupational therapists are excellent sources of information regarding aids that might help with eating, such as specially adapted cutlery or non-spill cups.

At the end of stage 6 on the FAST scale, those with Alzheimer's dementia begin to experience swallowing difficulties (called dysphagia). A speech and language therapist can be a valuable asset to assist in dealing with dysphagia. Symptoms of dysphagia include holding food in the mouth, continuous chewing, and leaving harder-to-chew foods such as hard vegetables on the plate. If the person with dementia is drowsy and/or is lying down, he or she might struggle to swallow safely. It is important to ensure that the person is alert, comfortable, and sitting upright or, if in bed, well positioned, before offering food and drink. Physical therapists can suggest positioning techniques. A referral from the doctor for certified home care services can provide access to a range of specialists including: nurses, occupational therapists, physical therapists, speech and language pathologists, and social workers. These services are provided under the Medicare benefit.

Providing enough fluid to those with Alzheimer's dementia is also important. The person should be encouraged to drink about 1.2 l of fluid throughout the day. It is important for the caregiver to encourage drinking and to actually watch to see if the person drinks what is provided. Placing a drink in front of someone does not assure that he or she will drink it. Also, an empty cup does not always mean that the person has consumed the contents. It may have been spilled, drunk by someone else, or poured away. Tips to ensure adequate hydration include:

- Have a drink immediately available whenever the person is eating something.
- Use a clear glass so the person can see what is inside, or a brightly colored cup to draw attention.
- If possible, offer the person the cup or put it in his/her line of sight.
- Describe what the drink is and where it is, so that if the person has a problem with sight they are still able to find the drink.
- Offer different types of drink (both hot and cold) throughout the day.
- Make sure the cup or glass is suitable—not too heavy or a difficult shape.
- Foods that are high in fluid can help, e.g., gravy, jelly, or ice cream.

Challenging Behaviors That Surround Eating

A person with Alzheimer's dementia may refuse to eat food or may spit it out. This may be because they dislike the food, are trying to communicate something such as the food being too hot, or they are unsure what to do with the food. The person may become angry, agitated, or exhibit challenging behaviors during mealtimes. These behaviors occur for a variety of reasons, such as frustration over difficulties they are having, feeling rushed, the environment is unsettling, the people they are eating with are disliked, the food is disliked, or they may not want assistance with eating. It can be a challenge to identify what the problem is, particularly if the person is struggling to find the words to explain. It is important to remember that these reactions are not a deliberate attempt to be "difficult," or a personal attack on the caregiver.

Those with Alzheimer's dementia have only one "speed" and trying to rush them along only leads to problems. It is important for caregivers to be attuned to the nonverbal behavior of the person. If the person appears to be agitated or distressed, this is not the time for a meal. It is essential that the caregiver(s) look for nonverbal clues such as body language and eye contact as a means of communication. It is much better to wait until the person is calm and less anxious before offering food and drink (http://www.alz.org/care/alzheimers-food-eating.asp, Accessed 1 June, 2014; Trivedi 2014).

Eating Environment

- Eat with the person. This will help make eating a social activity and can also help maintain independence as the person may be able to copy the actions of the caregiver.
- Make the environment as stimulating to the senses as possible. Familiar sounds of cooking, smells from the kitchen, and familiar sights such as tablecloths with flowers can all help.
- Some people enjoy eating with company; others prefer to eat by themselves. These choices might vary from one meal to another. Either way, it is important to make sure the person has enough space.
- A noisy environment can be distracting. The eating environment should be calm and relaxing.
- The person with dementia should be able to choose where she/he wants to sit and eat. They should also be able to choose what they want to eat, within reason.
- Some people with dementia will also have problems with their sight. They may not be able to see the food in front of them. It is important to make sure that the food is colorful and the environment is well lit.
- Use color to support the person—the colors of the food, plate, and table should be different. Avoiding patterned plates is important.
- Try not to worry about mess; it is more important for the person to eat than to be neat (Alz site-eating tips 2013).

Meal Preparation

It is important to keep those with dementia involved in preparing food and drink. This helps to maintain their skills in this area. The caregiver could break down preparation into individual tasks, for example, preparing the vegetables or buttering bread. The person with dementia may also help with the shopping, if accompanied. The caregiver may help the person with online ordering and delivery, which may be a way to ensure that there is fresh food in the house.

Living Alone

Meals on wheels may be an option for people in FAST scale stage 5 who are able to live independently, but need support to prepare food. However, the service is not available everywhere. Generally, churches are involved in this service. If not, the Alzheimer's Association can usually direct families to available resources in a specific area. Frozen or refrigerated ready meals are another option. The presence of spoiled or hidden food is a sign that the person with dementia may need extra support. It is possible to arrange for private duty home health care nursing assistants to visit the person at mealtimes and either prepare a meal, or remain in the home while the person eats. The local social services department can provide additional information regarding these services. Nursing assistants from certified home care agencies can also provide necessary help—however, such services are seldom provided on a daily basis and are usually available only intermittently and for a designated period of time (Lokvig 2003).

Managing Incontinence

The management of continence is directly linked to eating—what goes in is utilized by our bodies and turned into waste that must eliminated. At the end of stage 6 on the FAST scale, where the cognitive level is that of about a 2- to 3-year-old, those with dementia will start experiencing difficulty managing the mechanics of toileting, i.e., pulling pants down so that clothing does not get wet from urine or soiled from feces; lifting the seat up to urinate for men, putting the seat back down, and ensuring that urine is deposited into the toilet and not on the floor of the bathroom; pulling pants down for women, sitting back far enough on the toilet seat that urine and feces are deposited into the toilet, pulling enough toilet paper off the role to adequately wipe, flushing the toilet, and washing hands. The loss of these skills is a harbinger of complete incontinence. However, the person may not need to wear incontinence products just quite yet. In fact, putting someone into these products before he/she is truly incontinent hastens the incontinence and may lead to UTIs. Once someone is wearing these products the caregiver is less concerned with the person's voiding and defecating—there is no longer worry that waste products will be deposited on a chair, a sofa, or in a bed. So, the person may stay in a wet or soiled incontinence product for hours. When total incontinence is present, the person is nearing the end of the FAST scale at stage 6. At this point, it may be possible to manage incontinence but there is nothing that can be done to prevent these changes. They are part of the ravaging of the body that comes with Alzheimer's and other dementias.

So, what can be done to help manage incontinence? Again, avoiding placing someone in incontinence products too early is one strategy. The other is for the caregiver to set up a "bathroom-drinking" schedule that begins when the person starts his or her day. The first thing that all of us who are cognitively intact do in the morning is go to the bathroom and urinate. We get rid of all the urine that has collected in our bladders during the night. Then we might get a cup of coffee or tea to get us started for the day. Throughout the day we drink water, coffee, tea, and whatever other liquids we like—thereby keeping ourselves hydrated. We also urinate and defecate throughout the day as our bodies signal us to do so. But what happens to those who are no longer able to independently manage these basic skills? They probably remain in the wet diaper that they slept in for a period of time before they are changed—they also might start their days with coffee or some other beverage—all the time sitting in a wet and warm diaper. Because they do not ask for drinks and no longer have the ability to make their own coffee or tea, they might wait several hours after awakening to receive something to drink and to have their diapers changed. Is there a better environment for a UTI to develop? What needs to happen is that the caregiver establishes an every 2-h drinking and going to the bathroom schedule. This schedule helps to eliminate UTIs with the potential of delirium and also maintains the integrity of the person's skin.

This chapter ends with a story from Beverly Bigtree Murphy. Beverly took care of her husband Tom, who was diagnosed with Alzheimer's, and she cared for him until he died. Beverly now has a very helpful website that deals with many of the challenges of caring for a loved one with dementia. The following is from her website and it deals with how she helped her husband to manage his incontinence:

> He became completely incontinent by 1990. We were lucky again. The level of intimacy the two of us enjoyed with each other allowed an easier transition for us as his needs changed. The crossing point for both of us occurred when I realized he wasn't cleaning himself properly. I took the practical approach and offered what I saw as the only alternatives. Either he let me help, or someone else was going to have to help, or he'd end up dying of filth. Tom was pragmatic as always. As for my attitude towards doing 'it', I could see only four choices. I could choose anger, disgust, benign complacency, or love. I made the conscious effort to choose love. Assuring him I could also do this for him without revulsion or personal pain made the transition a lot easier for him and for me. Just to demonstrate that moments of tenderness were possible in what had to be thought of as an impossible breeding ground for such moments, I'll never forget the following incident.
>
> Tom had contracted a stomach virus which brought with it all the horror one can only imagine in the care of someone with the combination of motor control and reasoning problems Tom had. If there is hell on earth, diarrhea in a late stage Alzheimer's person is it. If there is a heaven, it only lasts 24 hours. Given my state of sleep deprivation and the physical energy it took to handle Tom's relentless need for help, I was at my wits end as we once more made the trip to the bathroom with pajamas that had to be changed and floors that had to be cleaned. I wanted to scream I felt so tired, and used, and spent, and unappreciated. And poor Tom, he was just responding to being sick and anxious from all the activity surrounding that night, getting more and more combative as the night wore on and as my own anxiety level increased. I remember steering him one more time into the bathroom and out of desperation, (I don't where the words came from,) I uttered.
>
> "Thank you, Lord!"
>
> And Tom responded.
>
> "Thank you for what?"
>
> "It could be worse, Tom."
>
> "How?"
>
> "I might not still love you as much as I do."

> He turned his head towards me and said.
> "You still love me?"
> "Of course, I love you."
> "God, I love you too." I saw tears form in his eyes and he continued.
> "Bevy!"
> "What, Tom."
> "Thank you."
> The evening became one of sharing the absurdity of what was going on in our lives and one of reconnecting to each other. And it was a sort of turning point for me. I stopped feeling victimized by what Tom's care needs meant by simply moving out of anger and into love, realizing in the process that it was as easy as making the choice to do so. I had two choices in attending to Tom's needs and that was to do them happily or do them mad. Either way, they still had to be done. And I realized something else, I realized how much Tom still needed to be loved. After all, what were we actually dealing with? A little poop between lovers wasn't really that much of a big deal. (Bigmurphy.com Moments of Love During Trying Times. Accessed July 11, 2014, http://bigtreemurphy.com/Using%20Usual%20 Things.htm. Do we need permission to quote Murphy's story??)

Beverly Bigtree Murphy generously shares with readers the struggles and more importantly the successful strategies that she learned as she provided care for her husband Tom. Many of the strategies we suggest are taken from Beverly Bigtree Murphy's website.

Eventually, everyone who is diagnosed with AD or another progressive dementia will become incontinent. First, the person becomes incontinent of urine and then of stool. Some will advise you that incontinence is a reason to place the person with dementia in a nursing home. This is not true. In fact, many people with dementia start to have problems with incontinence while they are still pretty involved in the world. Experiencing "accidents" is common in the early stages. Incontinence is not a reason for institutionalization—the caregiver can manage this and can develop the skills needed to keep her or his loved one at home as long as this is a feasible option.

Accidents will happen before the person is totally incontinent. It is important to be a hovering presence. During the daytime, it helps to encourage the person to use the bathroom about every 2 h. That, however, is not always possible. The incontinence products serve as a safety net when the toileting schedule is not always adhered to. If the schedule adds to the stress of caregiving, than ignore it—just make sure that drinks are frequently offered to the person with dementia—otherwise the consequence could be fecal impactions and UTIs. Remember that it is a lot easier to avoid these complications than to have to deal with them. One thing for sure, do not ask the person if she or he needs help—most likely the answer will be "no." Work from behind the person—this minimizes your presence. Treat incontinence pads in a matter of fact way just as if they are simply another part of the dressing routine. Avoid making statements like, "It's time to change your diaper." Instead, use terms such as "showering," "bathing," "getting freshened up," or "getting ready for bed."

Do not ask the person with dementia for permission to use incontinence pads—this is the caregiver's decision. Have all the supplies needed to "clean someone" at hand and organized. This will increase the ability to work efficiently, competently, and matter of factly. Leave the cleaning up of the room until after the person is dressed and in another area.

Additional Strategies for Dealing with Incontinence

1. Put a picture of a toilet or word toilet in large print on the bathroom door. Leave the door open.
2. Differentiate the toilet seat color. If the bathroom is all white for instance, get a colored toilet see so the person can distinguish where the toilet is and this will make it easier for them to make the toilet.
3. Stock up on inexpensive washcloths sold in packages of 6 or 12. They are thin and allow the caregiver to get adequately clean the individual's bottom and grab hold of feces. Use different colors of cloths for different functions. Baby wipes and similar products are not only expensive but they do not provide a good grip, which is important when removing feces. Have a stack of these cloths ready—some soaped and some ready to rinse. If using the packaged products, which offer convenience on the road, pull them out before cleansing the person. Some of these are flushable. If traveling look for highway gas stations which offer more privacy of a single stall plus access to a sink. Always travel with a dispenser of liquid soap for cleansing—using a mixture of ½ soap and ½ water for ease in dispensing, which at the same time does not put on too much soap. Use a dispenser with a pull top, since tab tops get in the way.
4. To deal with "diaper rash" that can quickly escalate without proper treatment. Use cold pressed castor oil or bag balm—both make good lubricants and salves for skin breakdown, rashes, etc.
5. Encourage your loved one to use the bathroom before bedtime. Do not awaken that person during the night to use the bathroom—not only will you be interfering with her or his sleep but yours as well. One of the major enemies to successful caregiving is the exhaustion of the caregiver. (Bigmurphy.com *Moments of Love During Trying Times, Accessed 11 July 2014,* http://bigtreemurphy.com/Using%20Usual%20Things.htm.)

Making the Bathroom Safe and Friendly

These suggestions are important for managing urinating and defecating but also for bathing and brushing teeth as well:

1. Cover or remove wall-to-wall mirrors—seeing the reflections in the mirror confuses the person with dementia, who might have trouble determining whether there are strangers in the room. The latter could precipitate a catastrophic reaction where the person and/or the caregiver get hurt. Mirrors on a medicine cabinet do not generate the same response.
2. Remove glass sliding doors—they are an accident waiting to happen.
3. Obtain a hand-held shower nozzle at least 7 ft long. Purchase a shower chair. When the person shows resistance to moving under the shower head to rinse off they are probably afraid of the shower (see our teaching tool on bathing and the

value of the tootsie roll pop in Chap. 10). This is when the hand-held shower nozzle will work wonders in allowing the caregiver to thoroughly rinse off the person.
4. Remove as much "stuff" from the bathroom as possible—the simpler the bathroom the better. Perhaps paint the bathroom in a solid soothing color, with few distractions on the walls or on flat surfaces. This goes a long way to help make the bathing experience one that is free from agitation and aggressive behavior.
5. As noted above, when traveling by car, it is better to handle incontinence issues in gas station bathrooms rather than large multi-stall bathrooms designed for either women or men—but not both. The bathroom in a service station is usually one room with a toilet and a sink. This allows the activities involved in cleaning someone's bottom to be done privately (Bigtree Murphy 1996).

Summary

When AD or another dementia develops, this changes everything—it rolls back the cognitive and functional abilities that most people not only accept as the norm but also take for granted. It is difficult for most persons to envision that a loved one could require the level of care that is described in this chapter. Deciding what and when a person eats and all the details that surround eating—shopping for food, putting food away in refrigerators and pantries, deciding on what to eat and actually preparing the meal—these are all activities that are usually taken for granted. Even more basic, is the ability to manage waste products—not only to go to the bathroom when the urge is felt but also to handle all the activities that accompany voiding and defecating—pulling pants down and up, sitting on or standing in front of the toilet, wiping, washing hands—these are such old skills that seldom even stimulate a thought. Most automatically assume that these skills will be theirs forever, but what if.....?

References

Alz site-eating tips (2013). http://www.alzheimersreadingroom.com/2013/01/how-to-get-alzheimers-patient-to-eat.html. Accessed 01 June 2013.

Bigtree Murphy, B. (1996). *He used to be somebody, 1995: A journey into Alzheimer's disease through the eyes of a caregiver.* Gibbs Associates.

Carson, V. B. (2013). Eating, appetite, and Alzheimer's. *Caring, 32(10)* 48–49.

Lokvig, J. (2003). *Alzheimer's A to Z: Secrets to successful caregiving.* Endless Circle Press. http://www.parapublishing.com/sites/para/resources/successstories_detail.cfm

Increasing links on food and depression. Understanding nutrition, depression and mental illnesses. http://www.ncbi.nlm.nih.gov/

Trivedi, M. H. (2014). The Link Between Depression and Physical Symptoms. http://www.nia.nih.gov/alzheimers/features/promoting-successful-eating-long-term-care-relationships-residents-are-key. Accessed 03 June 2014.

http://ajcn.nutrition.org/content/71/2/637s.full.

Chapter 7
Sexuality and Intimacy in Those with Dementia

Another issue that challenges family and paid caregivers alike has to do with the expression of sexuality in older adults with dementia. Contrary to common perceptions, sexual behavior may actually increase in those who have dementia. Even with profound cognitive impairments, many of those with dementia enjoy a high degree of sociability and the capacity to engage in intimate relationships. In fact, to be human is to be sexual—we are "hardwired" for this. Sometimes, the desire to be close to another person manifests itself in positive feelings of being cared for and desirable. Sometimes, this same desire for closeness is manifested in flirtation, affection, passing compliments, proximity to another, as well as sexual intercourse. The expression of sexual feelings can be directed to a spouse, a professional caregiver, to other residents in a facility, as well as toward adult children of the person with dementia. The adult child might closely resemble the person's spouse when at a similar age. In this case, the sexual overtures are basically a case of "mistaken identity."

None of us ever outgrow the need to touch others and to be touched, which is deeply ingrained in our biology. However, as we age, we are touched less and those with dementia are probably touched the least. This lack of touch might be a factor in the sexual behavior that is seen in those with dementia (Heerema 2013)

In general, there is a reduction in sexual drive in about 25 % of those with dementia (Miller et al. 1995). An increase in libido is reported in about 14 %. The incidence of sexually inappropriate behavior is reported to be relatively low in persons with dementia—ranging from 1.8 to 15 % in samples of residents in assisted living facilities (ALFs) who are diagnosed with dementia (Tsai et al. 1999; Alagiakrishnan et al. 2005). However, for a family or caregiver who is confronted with challenging sexual behavior, the low percentage is meaningless—to the recipient of the inappropriate behavior, it feels like 100 %!

A survey of 250 residents in 15 Texas nursing homes found that 8 % said they had sexual intercourse in the preceding month and 17 % more wished that they had. In the journal *Clinical Geriatrics,* 90 % of 63 physically dependent nursing home residents said they had sexual thoughts, fantasies, and dreams (Cirillo 2014).

V. Benner Carson et al., *Care Giving for Alzheimer's Disease,*
DOI 10.1007/978-1-4939-2407-3_7

Some of the concerns that surround sexual activity in ALFs and skilled nursing facilities (SNFs) are the occurrence of nonconsensual sexual activity, unwanted sexual comments, advances, coercive touch, sexual inhibition, including exposure of genitals, public masturbation, nudity, and hypersexual desire/demands, the use of obscenities, and false accusations of sexual abuse between residents as well as between staff and residents.

Sexuality is not determined by locale and continues wherever the person with dementia resides. These facilities are scrambling to develop policies to provide guidance and legal security as the incidence of sexual behavior among residents is increasing.

Sexuality is one of the major issues of conflict for family members as well as for caregivers within ALFs and SNFs. Both families and caregivers generally support the individual resident's right to choose, but when confronted with sexual activity, the "right to choose" comes into question. One of the issues is the difficulty for staff members who work in a senior living facility to support decisions that do not align with their own beliefs; likewise, adult children may express displeasure with choices their parents are making that may seem out of character.

Risks Associated with Sexual Activity Among Older Adults

The frequency of sexual behavior in older people, with or without dementia, is significant. Fifty percent of those between the ages of 65 and 74 are sexually active, and 25 % of those aged 75 and older are sexually active. Only 38 % of men and 22 % of women over age 50 discuss sexual problems with their primary care physician (Bancroft 2007).

This fact is important for a number of reasons. Once the risk of an unwanted pregnancy no longer exists, individuals cease using protection when engaging in sexual intercourse. Old age does not lessen the risk of contracting and passing on a sexually transmitted disease (STD). In fact, a 2013 report from the Centers for Disease Control and Prevention (CDC) stated that the incidence of STDs among older Americans is on the rise (CDC 2013).

An estimated 25 % of Americans who are HIV positive are over 50 years of age. Chlamydia rates among men between ages 45 and 64 increased nearly 200 % between 1996 and the end of 2006, similar to that in women. The reason for this increase is that a significant number of men in that age group remain sexually active, some into their nineties, due to the use of erectile dysfunction drugs such as Viagra, Levitra, and Cialis. This same study by the CDC showed that STD rates among men using Viagra were twice those of males not using the drug. A study in the *Annals of Internal Medicine* found that older men who use Viagra and similar drugs are six times less likely to use condoms compared with men in their twenties (Jena et al. 2010). Furthermore, STD educational programs seldom target the elderly.

Increasingly, ALFs and SNFs are requiring STD screening and a sexual history as part of the required medical examination prior to admission. The following case study illustrates the need for STD screening.

Sam, a 70 year old, was recently been admitted to an Assisted Living Facility. Soon after admission, Sam developed several lesions on his penis and inner thighs. The nursing staff concluded that the condition was due to incontinence and the use of incontinence pads. The nurses treated the lesions with a protective skin barrier, but there was no improvement. Only after reviewing Sam's medical records and consulting with the physician did they realize the truth. Sam had contracted genital herpes almost twenty years earlier and still suffered from periodic outbreaks. In this case, proper treatment, infection control, and staff education were delayed because the nursing staff had failed to thoroughly review Sam's medical history and failed to recognize a common sexually transmitted disease.

Additional Facts About Sexuality Among Elderly with Dementia: The Value of Touch

Not everyone diagnosed with Alzheimer's will have the same interest in sexual activity. For some, interest in sexual activity wanes in the earliest stage of the disease. Others continue to enjoy sexual activity through the early part of stage 6 on the functional assessment scale (FAST) scale. However, whether or not the person with Alzheimer's continues to be sexually active, that person benefits from touch. Touch decreases sensory deprivation, increases reality orientation, stimulates the mind, decreases pain, isolation, and vulnerability, and conveys trust, hope, and reassurance. The person with dementia may still enjoy being stroked and hugged but may be unable to initiate such physical affection. When partners no longer share a bed, some people with dementia find it comforting to have something to cuddle, such as a soft toy or hot-water bottle.

A 1-year study looked at the effects of gentle massage on two groups of older nursing home residents. One group was suffering from chronic pain, and the second group had dementia and was exhibiting anxious and/or agitated behaviors. The certified nursing attendants were trained by a licensed massage therapist to deliver "tender touch" massage. The project was divided into three 12-week phases, where different staff and residents were involved in each phase. Fifty-nine of 71 residents completed the 12-week program. Pain scores declined at the end of each phase, and anxiety scores declined in two of the three phases. Eighty-four percent of the nursing attendants reported that the residents enjoyed receiving tender touch, and 71 % thought this type of massage improved their ability to communicate with the residents (Sansone and Schmitt 2000)

Increased Sexual Interest

Some people with dementia find that their desire for sex increases. Some partners find this a welcome change, while others feel unable to meet the level of sexual demand. When the latter is the case, it can be difficult for the person with dementia. In this situation, some partners have said they feel wary of showing normal affection in case it is mistaken for a sexual advance. If the level of sexual demand feels overwhelming, it can be helpful to find something else to do together that can meet the other person's need for intimacy, rather than making an outright refusal to a sexual overture.

Sometimes, aggressive behavior is a response to a sexual overture being hurtfully refused. It helps if the person who is turning down the overture considers the feelings of the person who is initiating the sexual encounter. A response that acknowledges the initiator's needs, is respectful, and does not hurt the feelings of the other may help to prevent aggression. In some cases, it may be a good idea to keep safely out of the sexual initiator's way until his/her mood has passed.

When demands for sex are inappropriate—either directed to the wrong partner or demanded too often, it is important to speak about this with the person's physician who may prescribe medication to lessen sexually aggressive behavior. If the person is behaving in a way that distresses private duty nursing assistants when they are providing personal care—for example, when they are providing help with bathing, toileting, and dressing—family members can become embarrassed and may feel they should stop employing outside help. It is important that families share the concerns of the care workers with the physician, who may be able to make suggestions about strategies to handle the situation. For instance, it is normal for caregivers to giggle or laugh from embarrassment when a person with dementia makes sexual comments or tries to touch the caregiver in an inappropriate manner. The best response is a consistent one from each caregiver. If the caregivers all respond in the same way, the inappropriate behavior may diminish. The response should include a stern face without a smile, and a firm verbal response such as the following: "Mr. Jones, I do not like it when you talk to me like that—it is wrong!" There can be no smiling or giggling on the part of the caregiver, only firm and consistent limit setting will help this situation. If that strategy does not work, the family caregiver needs to discuss the behavior with the physician who might prescribe medication that can help (such as citalopram 40 mg/day).

Challenging Sexual Behavior

For many couples, where one of the partners has Alzheimer's, sexual relations carry on as usual. Others say that their partner with dementia can appear cold and detached during sex. Still others report that shortly after sexual relations, the person with dementia forgets that he/she has engaged in sexual relations and almost immediately demands to engage in sex again. Sometimes, the person with dementia no longer seems to recognize his/her partner. These situations can be upsetting and painful for the nondemented partner.

If the person mistakes someone else for her or his partner, it is important that the mistaken "partner" try to approach the situation in a manner that as much as possible maintains the dignity of the person with dementia. This means avoiding accusations or reacting in a horrified way. The situation is uncomfortable for both the person with dementia and for the person who has been mistakenly identified as a sexual partner. Talking calmly and privately reduces the potential for the demented person to feel embarrassed or distressed.

In some situations, although infrequent, people with dementia may go through a phase of being sexually aggressive—making repeated demands for sex from her/his partner or other people. In extreme cases, particularly if the person with dementia is a strong person, the threat of physical force may be difficult to manage. For some people, this behavior may be part of a long history of aggression, which is worsened by the dementia. When this happens, again, the issue must be shared with the person's physician.

It is important for the cognitively intact partner to try to avoid responding to the person's behavior as if the behavior was deliberately intended to be hurtful or embarrassing. Challenging behaviors of all kinds are caused by brain damage and are not conscious choices of the person with dementia. However, partners and caregivers must always be cognizant of their own safety. If either feels pressured to participate in sexual activity, or feels threatened, or is the target of aggression or verbal abuse, this situation must be addressed medically.

Changes in Level of Inhibitions

For most people, sex is a very private matter, and many people find it difficult to talk about their sexual feelings even to their partners—let alone anyone else. However, living with dementia brings all kinds of private issues into the public arena.

Dementia can reduce a person's inhibitions due to damage to the frontal lobe of the brain, thus causing people to publicly express private thoughts, feelings, and behaviors—including those relating to sex. The loss of inhibitions may lead to sexual advances to others, undressing, touching, or masturbating in public. The person with dementia may use crude and offensive language—behaviors that prior to developing Alzheimer's disease were completely out of character for this person. These situations may be embarrassing for those close to the person, but they may also be very confusing, distressing, or frustrating for the person himself or herself—especially if the person cannot understand why the behavior is considered inappropriate. This kind of behavior rarely involves sexual arousal. Sometimes, what appears to be sexual is actually an indication of something quite different, such as:

- Needing to use the toilet
- Discomfort caused by itchy or tight clothes or feeling too hot
- Boredom or agitation
- Expressing a need to be touched, or for affection
- Misunderstanding other people's needs or behavior
- Mistaking someone for their current (or previous) partner

It is common for those who are emotionally connected with the person with Alzheimer's to want to protect the person with dementia from others laughing at them or from being shocked by her or his behavior. The response to embarrassing behavior might be to ask certain people, such as grandchildren, not to visit. Decisions like this add to the isolation often experienced by the primary caregiver. The situation is another one that should be shared with the overseeing physician who might prescribe a medication to help.

Adapting to Changes in a Partner's Behavior

The partners of people with dementia describe a wide range of feelings about their continuing sexual relationships, ranging from pleasure that sex is something that they can still share, to confusion at being touched by someone who at times seems like a stranger. As dementia progresses, the situation often changes, as do the feelings of those involved.

On the one hand, partners' feelings may not change towards the person they are caring for at all—they may find that they can connect with their partner through sex even if they are finding it difficult to communicate in other ways. On the other hand, some caregiving partners feel exhausted by their responsibilities and report that they do not have the energy to enjoy sex. This can be frustrating for their partner with dementia. Some partners find that the intimate tasks they have to perform for the person with dementia discourage them from engaging in sexual activity. The lack of sexual interest coming from the caregiver can make a person with dementia feel that he/she has lost sexual appeal, thus leading to negative feelings about self as well as about his or her partner.

Many caregivers find it hard to enjoy a sexual relationship if other aspects of the relationship have changed. If nothing else is shared between them, it may seem that sex too is meaningless. If this is the case, the cognitively intact person needs to provide the demented partner with plenty of reassurance and affection in other ways.

Depending on how the dementia affects their relationship, some partners continue to sleep in the same bed as their partner with Alzheimer's. Others choose to move to single beds or separate rooms. If a partner does decide to move into another room, this can be disorienting or distressing for the person with dementia, so it is important to discuss this with the overseeing physician.

When the Caregiver or Person with Dementia Feels Frustrated

In any relationship, problems can arise when one person wants to have sex and the other does not. This is a situation that most people in long-term relationships go through at one time or another, and it is important to remember that it can arise even when dementia is not involved. The partners need reassurance that this is normal. Single people also have sexual needs and may become frustrated when these are not met. If a family member or a friend is arranging the personal or household care for someone who is living alone, it is a good idea to talk regularly with the care provider(s) to see if there are any behaviors that they are finding difficult. For a variety of reasons, care providers may find inappropriate sexual behavior difficult to mention. However, it is important to know about their experiences and to validate their feelings.

What About Sexuality in an Assisted Living or Skilled Nursing Facility?

Some of the concerns that surround sexual activity in ALFs and SNFs are the occurrence of nonconsensual sexual activity, unwanted sexual comments, advances, coercive touch, sexual inhibition, including exposure of genitals, public masturbation, nudity, and hypersexual desire/demands, the use of obscenities, and false accusations of sexual abuse between residents as well as between staff and residents.

Sexuality is not determined by locale and continues wherever the person with dementia resides. These facilities are scrambling to develop policies to provide guidance and legal security as the incidence of sexual behavior among residents is increasing.

Increasingly, ALF administrators are approaching the issue of sexuality with sexual intercourse is just one of the concerns facing ALF and SNF administrators. As mentioned earlier, there are a range of sexual behaviors that can cause concern. Generally speaking, ALFs hold to a philosophy of honoring the resident's wishes; however, the question of whether or not the resident is cognitively able to know what is happening makes it difficult to adhere to that philosophy. What if the resident with dementia wants to engage in sexual activity with someone who does not have any memory impairments? The provider is then faced with the challenge of verifying that the resident with dementia is aware of both the risks and benefits of her/his actions, is able to say "no" at any time, and is not at risk of being exploited.

The issues related to sexual activity within a facility include:

1. The need for a policy that addresses sexuality between residents. Such a policy should be shared with incoming residents as well as family members and should be a critical piece of the orientation for new employees (see sample policy from Hebrew Home in NY in the Appendix and Guidelines in this chapter).

2. Training for all staff on how to respond to a variety of sexual scenarios, including being the target for sexual comments and unwanted touches, interrupting a male and female couple engaged in sexual intercourse, interrupting a same-sex couple engaged in sexual intercourse, witnessing a resident publicly masturbating, and so forth.
3. A process to determine the competency of residents to make sexual decisions that affect themselves and others.
4. A plan for managing resident's rights regarding sexual activity while protecting the rights of others.
5. Family involvement in decisions involving the sexual behavior of a resident.
6. Strategies to deal with reactions of other residents to sexual behavior that they witness.
7. Documentation of sexual behavior and communication of this behavior to overseeing physician and family.
8. Role of other providers regarding sexual activity in the facility, including administrators, the physician overseeing the facility, the resident's primary care physician, nursing staff, ombudsmen, and others.

Staff Reactions to Sexual Behavior

For many staff providers, this is an area that reaches beyond their comfort zone. When they are confronted with sexual behavior, they may react with a variety of emotions, including embarrassment, confusion, and a sense of inadequacy regarding how they should respond. Training that includes strategies on how to respond to a variety of sexual scenarios (including role-playing) is vital. Staff members need guidance regarding how to talk to elderly patients and their families about sexuality and sexual behavior. Most likely, the elderly person has experienced sexual activity and that experience is stored in her or his long-term memory. However, caregivers have no way of knowing whether those memories are pleasurable or painful. Only the resident's response in the present situation provides a hint to the nature of those memories.

Issues for Families

The initial response of most family members to a loved one's sexual activities is embarrassment—"Not my mother!" Shock is a very common reaction. Children of all ages have difficulty viewing their parents as sexual beings, so the idea that a mother or father could be having sex with a strange woman or man is invariably upsetting. Many family members will believe that this behavior is morally wrong. Family members frequently will make unreasonable demands of the facility staff to prevent any sexual liaisons, and may even threaten to sue the facility. Other family members, most likely in the minority, are supportive to whatever makes their mother or father happy.

Assessing Competency

The ability to assess whether or not an elderly person is competent to engage in sexual relations is not an easy task, especially if the person has dementia. Asking certain questions will help to determine competency. For instance, is the individual aware of the relationship? Does he or she know who is initiating sexual contact? Can she or he describe a preferred degree of intimacy? Is the person capable of avoiding exploitation? Can he/she say "no"? Is the behavior consistent with formerly held values? Is the person aware of possible risks? Does he/she realize the relationship maybe temporary? Can he/she describe how she or he will react when the relationship ends?

Determining Consent and Safety

There is no easy answer to determining consent, and it may be different for each person with dementia. The facility must have policies and guidelines established that address this issue. These guidelines provide a basis upon which to make decisions regarding what is appropriate sexual behavior. It is important that a sexual history is obtained as part of the admission interview prior to entry into a facility. Each resident is entitled to his or her right to seek out and engage in consensual sexual relations, the right to privacy, and the freedom to make decisions. However, there is an old saying "Your rights end where my nose begins!" This means that one resident cannot force another resident to engage in sexual behavior or any other activity that they do not freely consent to.

What constitutes consent? The person needs to be able to:

- Understand the information provided.
- Retain that information long enough to make a decision.
- Weigh available information in order to make a decision.
- Communicate that decision via language, using sign language, or even simple muscle movements such as blinking an eye, or squeezing a hand.

Sometimes, decisions about sexuality are made by a family member or a good friend. It is not that desires for sexual expression go away. Rather, what disappears is the ability to consent to how and with whom these needs are expressed.

Guidelines

Sexual expression should be permitted if both parties and relevant family members consent and the risks are not judged to exceed the benefits to those impacted by this decision. Staff members are responsible for determining and documenting consent, and for developing a plan of care. It is in everyone's interest and the staff's responsibility to seek a mutually agreeable solution when family members object to consensual sexual behavior between residents. The following case illustrates the difficulties when there is lack of agreement between family and staff.

> Mr. Blake's daughter came to visit him in the ALF where he lived. When she arrived at his room, she was unable to open his door. He had it barricaded from the inside. The daughter could hear talking and laughter from inside the room, but her father did not open the door. The daughter looked for a staff member to assist her to open her father's door. The staff member was unable to do so. Despite the loud knocking and calls to her father from the outside, he did not respond. After about 30 min, her father opened the door and the daughter saw a female resident in her father's room. The woman looked disheveled, her hair was not combed, and her blouse was open. The daughter became very upset. She was sure she had interrupted her father and the woman who had probably been engaging in sexual activity. The daughter found the administrator of the ALF and told him that her father was having a sexual relationship with a woman who was not her mother and this behavior blemished the memory of the 50 years her mother and father had been married. The daughter demanded that the administrator move the woman to another floor of the facility. This incident prompted an interdisciplinary conference, including nursing staff, the attending physician, the psychiatric nurse practitioner, the daughter, and the ombudsman. After much heated discussion, the clinical staff concluded that the two residents were competent to make their own decision regarding whether or not to engage in sexual activity. The daughter was very angry with that decision and decided to move her father to a VA Assisted Living Facility where there were only male residents.

As noted earlier, it is essential that both ALFs and SNFs assess intimacy/sexual needs at the time of intake, and that both families and staff members receive education about sexuality. As noted previously, agencies need to develop and put into writing their intimacy and sexuality philosophy, policies, and procedures. Also, as noted previously, facility staff needs to decide whether to obtain a sexual history from residents who are able to supply this information.

Inappropriate Touching

It is important that caregivers dress and behave modestly and do nothing to "tempt" a resident who might misinterpret cues. As noted previously, the use of a firm response, "Mr. Smith, you cannot touch me like that" as the caregiver firmly removes the offending hand may be sufficient. Also walking away from the person whose behavior is inappropriate frequently works to send a message regarding the inappropriateness of certain behaviors.

It is important for staff members and family to review whether there are specific times during the day or during certain scheduled activities when inappropriate behavior is more likely to occur. For instance, during dressing and bathing, it is not unusual for the resident to make inappropriate comments to the caregiver, touch inappropriately, and/or openly masturbate. The caregiver needs to perform these tasks in a "matter-of-fact" manner. An intervention such as switching a female caregiver for a male caregiver and vice versa also helps to diminish this behavior.

Summary

Sexuality is not something that is turned off because someone has dementia. The needs for touch, closeness, and affection continue until death. However, when someone with dementia seeks to fulfill his/her sexual needs, this can be a challenging issue for both family members and paid caregivers. Some caregivers are uncomfortable with the idea of sexuality and find it difficult to discuss this very private area. Other caregivers are more open in their attitudes regarding sexuality and may approach the sexual activity of a person with dementia with a "live and let live" attitude. These two approaches are miles apart. Family caregivers need to be able to discuss these issues with a health professional—an overseeing physician, a charge nurse, an administrator, or possibly an occupational therapist. If the person resides in a facility, then issues of sexuality need to be addressed before the persons is admitted to the facility. These are easily approached when one of the admission criteria is a screening test for STDs. Approaches and decisions made regarding how to address issues that surround sexual expression must consider not only the feelings and values of the person with dementia but also the feelings and values of their caregivers.

References

Alagiakrishnan, K., Lim, D., Brahim, A., et al. (2005). Sexually inappropriate behavior in demented elderly people. *Postgraduate Medical Journal, 81*(957), 463–466.

Bancroft, J. (2007). Sex and aging. *The New England Journal of Medicine, 357*(8), 820–822.

CDC Report. (2013). National report, incidence, prevalence, and cost of sexually transmitted infections in the United States. (13 Feb 2013).

Cirillo, A. (2014). Sex in nursing homes. *About.Com Assisted Living Where Do You Stand?* Accessed 30 May 2014.

Heerema, E. (November 2013). The importance of touch for people with dementia, *About.Com Alzheimer's/Dementia.* Accessed 29 May 2014.

Jena, A. B., Goldman, D. P., Kamdar, A., Lakdawalla, D. N., Lu, Y. (2010). Sexually transmitted diseases among users of erectile dysfunction drugs: Analysis of claims data. *Annals of Internal Medicine, 153,* 1–7.

Miller, B. L., Darby, A. L., Swartz, J. R., Yener, G. G., Mena, I. (1995). Dietary changes, compulsions and sexual behavior in Frontotemporal degeneration. *Dementia, 6*(4), 195–199.

Sansone, P., & Schmitt, L. (2000). Providing tender touch massage to elderly nursing home residents: A demonstration project. *Geriatric Nursing, 21,* 303–308.

Tsai, S. J., Hwang, J. P., Yang, C. H., Liu, K. M., Lirng, J. F. (1999). Inappropriate sexual behaviors in dementia: A preliminary report. *Alzheimer's Disease and Associated Disorders, 13*(1), 60–62.

Chapter 8
Communication Commandments

> Give us time to speak, wait for us to search around that untidy heap on the floor of the brain for the word we want to use. Try not to finish our sentences. Just listen and don't let us feel embarrassed if we lose the thread of what we want to say. Christine Bryden, diagnosed in 1995 at age 45 with Alzheimer's. (www.alzheimers.org.au/…/managing-changes-in-communication.aspx. Accessed May 14th, 2014)

The above quote is an expression of frustration felt by someone with early onset dementia and probably in stage 4 on the FAST scale. Christine is still alive and speaking around the world about her experiences with this disease. However, this quote emphasizes the importance of how we communicate to and with those who have Alzheimer's or other type of dementia. Christine was painfully aware of her faltering communication skills and was pleading for patience from her listeners. This self-awareness fades as the dementia progresses and the decline of communication skills marches on until only the individual's face, eyes, and muscle tone are left to speak for the person.

The Importance of Communication

Communication is vital to everything that individuals do. The accomplishment of tasks, the way in which individuals teach and learn, how the world is discovered, how relationships are formed and maintained, and individuals know what they know. All of this depends on the ability to for individuals to communicate to others and to understand what others are communicating to them. Alzheimer's disease (AD) and other progressive dementias slowly but surely deteriorates the fundamental skills of communication—not only does the individual lose verbal communication abilities so that expressing personal needs and feelings becomes difficult but also the ability to understand verbal communication directed towards the individual also diminishes. This breakdown in communication is an incredibly difficult and emotionally painful loss to accept and to manage. It leads to frustration not only for the person with this awful disease but also for caregivers. This chapter lays out

V. Benner Carson et al., *Care Giving for Alzheimer's Disease*,
DOI 10.1007/978-1-4939-2407-3_8

some specific and concrete strategies that facilitate communication across all stages of decline—from stage 4 on the FAST scale through the end of stage 7 which marks the end of the disease and death.

What Is Communication?

Words make up only a portion of communication with nonverbal communication making up the rest of what is "said" to others. Body language, tone of voice, gestures, facial expressions, touch, and eye contact or lack thereof all communicate. The implication of this is that when verbal communication deteriorates as a result of dementia, the caregiver needs to be increasingly aware not only of his or her own nonverbal communication but must also be equally tuned into the nonverbal communication of the person being cared for.

The skills of validating the meaning of both verbal and nonverbal communication are essential in providing care. Without validation, both the person with dementia as well as the caregiver will experience frustration, anger, and may withdraw. When verbal communication is diminished, validation allows effective communication to continue. The caregiver needs to put into words what he or she is sensing from the cared for person, while at the same time the caregiver must control his or her nonverbal messages so as not to communicate impatience, anger, disgust or other negative emotions. This is indeed a tall order. For example, a caregiver might validate in the following ways:

> Mom you get frustrated when you are not able to let me know what you need. I am sorry, let's try again.
>
> Honey you are tense, you are not smiling, and I notice your face looks sad to me when you get out of a chair. I wonder if you are in pain.
>
> Dad it must be frustrating to you that you need me to help bathe you. I understand and will try to make this the best experience I can for you.

The caregiver has the ability to validate; the person with dementia does not have this ability. Lacking this ability, the person will internally experience the tension in the caregiver and begin to feel the caregiver is unhappy and/or angry with them. This puts a great deal of responsibility on caregivers to provide not only gentle physical care but also care that is matched with kind and understanding words.

Deterioration in Communication Skills

Just as the disease of Alzheimer's produces slow but steady declines in all abilities that we associate with being an adult, the decline in communication skills follows a similar pattern of slow but steady deterioration. In stage 4 of the FAST scale, when people with dementia are seen as the "great foolers" and functioning between the cognitive age of 8 and 12 years, they have ability to act as if verbal messages are heard and correctly interpreted. When it becomes clear that the message was not heard or was misinterpreted, they have the ability to "cover up" their lack of understanding.

During stage 5 on the FAST scale, cognitive and functional abilities deteriorate from the level of a 7-year-old to that of a 5-year-old. The person can still make their needs known but their communication skills are simpler. They tend to use shorter words and they generally do not engage in long and involved conversations. During this stage, the person experiences word-finding problems and may engage in confabulation, that is, "filling in the blanks" with an answer when her or his memory fails. Confabulation, an unconscious response, is not to be confused with lying, a conscious attempt to deceive. During this stage, individuals with dementia neither use complex sentences nor understand complex messages. Communication, then, must be concrete and simple.

In stage 6, when the cognitive and functional level is that of a toddler from 4 deteriorating to 2 years of age, short-term memory has deteriorated to about 5 min so that messages spoken to the person literally go "in one ear and out the other." Messages must be straightforward and may need to be repeated before the individual can understand and respond.

During stage 7, the last stage, when functional and cognitive level is somewhere between that of an 18-month-old baby and a newborn, communication is initially limited to six or seven words and deteriorates until the person can no longer speak. Communication is accomplished through simple verbal messages, i.e., "I love you."; "Do you hurt?", "I hope this makes you feel good," and nonverbally through the senses. The person in the end stage responds to gentle touch, singing, soft music, wonderful smells wafting through the air, massage, and the presence of loving caregivers. These are the same expressions of love that we use with very young toddlers and infants.

Using the Ability to Read

Many caregivers are unaware that those with dementia can still read. The ability to read is another powerful and effective way to communicate. Unless the person is illiterate, reading is not only a very old memory, dating back to about 6 years of age or even younger, it is an "overlearned" skill. Those with Alzheimer's or other dementia may have started reading in the first grade of elementary school and every additional year of schooling required more reading—not only more in volume but the reading became more challenging, harder and required more thought to understand, remember, and apply. As an "overlearned" skill, reading is slow to be erased by the ravages of dementia. The person may only have 5 min of short-term memory but written instructions, labeled drawers, and written cues regarding activities of daily living allow the person to be more independent—not because he or she remembers what is to be done next but because they can read the labels on drawers that might say, "Put Underwear on First; Socks Second, Shirt Third and Pants Fourth."

Michelle Bourgeois, Ph.D., a speech therapist who is currently a professor at Florida State University, conducted ground-breaking research regarding the value of using the written word with those who have AD or related dementias. Bourgeois' work grew out of her Ph.D. research in the 1980s, when she developed some of the first memory books, which use pictures and sentences to help people with memory problems—including those with dementia—recall past events. Bourgeois contends that even when dementia is so advanced that people cannot speak, they can read if the words are large enough. She states that we know that they can read because they smile, make pleasant sounds, and stroke photos of loved ones with captions. A woman who attended one of Bourgeois' lectures reported that her father would repeatedly ask, "Where are we going?" during their weekly drives to the doctor. Bourgeois advised her to answer his question—and also write it down on a notepad and give it to him. When he asked again, she should say gently, "The answer is on that notepad." When the woman tried this out, she said that her dad looked at the notepad, out the window, and back at the notepad. After that, he stopped asking, "Where are we going?"

Similar techniques have been used to deal with anger and anxiety in people with dementia. When a patient refused to shower, Bourgeois told her nursing aide to make a card that read, "Showers make me feel fresh and clean" and give it to her after saying it was time to shower. The technique worked.

With a grant from the Alzheimer's Association, Bourgeois is working to dispel the belief that Alzheimer's makes people miserable. Using pictures with captions, she is asking patients about their quality of life. "We find that if caregivers aren't stressed and in a hurry, if the patient is well cared for, and if they feel safe and in a good environment, they think their lives are good," she says.

http://www.alz.org/research/alzheimers_grants/for_researchers/overview-2008.asp?grants=2008bourgeois Bourgeois has taught thousands of caregivers her methods, and they have taught thousands more. When she discovered over 20 years ago that memory could be reclaimed with simple tools, she set herself a high goal—one she still holds: "I want families to remember these as happy times in their lives" (Wicker 2010).

Effective Communication Strategies

When communicating with those with Alzheimer's or other progressive dementias there are a number of effective communication strategies to guide the caregiver including:

- Take a cleansing breath and consciously try to relax. This helps the conversation. It is important for the person initiating the conversation to have a topic in mind or a specific goal in mind. Is the purpose of the conversation just to connect in a loving way to the person or is there another reason?

- It is important for the speaker to stand close enough so the person is able to clearly see the speaker.
- Making eye contact is always a useful way to help the person focus on the speaker.
- Make sure that competing noises such as a radio, television, or other people's conversations are minimized.
- Use humor to enhance the conversation and relieve pressure. Laughing together is a great way to slide over misunderstandings and mistakes.
- Avoid showing impatience or annoyance towards the person—the ability to sense the "emotional climate" is a retained ability. However, the person may lack the ability to validate both verbal and nonverbal communications.
- Include the person in conversations with others. Being included in social groups can help those with dementia to preserve their sense of identity, while reducing feelings of exclusion and isolation.

What to Say

- Be positive. Remember that negativity is contagious.
- Do not ask too many direct questions. This can lead to frustration if the person does not know the answer. It helps if questions can be answered with a "yes" or a "no". Complicated decisions are beyond a person with dementia's capacity. Try not to ask the person to make complicated decisions.
- If the person does not understand the message that is being communicated, rephrase the message. Using the written word to supplement the spoken question can facilitate understanding of the message.
- As dementia progresses, the person may become confused about what is true and not true. If the person says something that the caregiver knows to be incorrect, rather than directly challenging the person, it is better to find ways of steering the conversation around the subject rather than directly confronting or correcting the person. If the content of what is shared is not "true," it is usually still possible to focus on the meaning and/or feelings they are sharing.

Listening

- Careful listening is always important. Sometimes the caregiver needs to encourage the person to continue to communicate. Listening carefully to what the person is saying, and providing him or her with plenty of encouragement keeps the lines of communication open.
- If the caregiver has not understood fully, it helps to repeat back to the person what the caregiver thinks she or he heard and ask for validation from the person with dementia. If the person has difficulty finding the right word or finishing a

sentence, ask them to explain it in a different way. Listen for clues. Also, pay attention to their body language. The expression on their face and the way they hold themselves and move about can give you clear signals about how they are feeling.

- If the person is feeling sad, let them express their feelings without trying to "jolly them along". Sometimes the best thing to do is to just listen, and show that you care.
- Due to memory loss, some people will not remember things such as their medical history, family, and friends. You will need to use your judgment and act appropriately around what they have said. For example, they might say that they have just eaten when you know they have not.

Body Language and Physical Contact

- A person with dementia can read body language. Sudden movements or a tense facial expression may cause the person to become upset or distressed and can make communication more difficult.
- Make sure that the body language and facial expression of the person with dementia match the shared communication.
- Never stand too close or stand over someone to communicate: it can feel intimidating. Instead, respect the person's personal space and drop below their eye level. This will help the person to feel more in control of the situation.
- Use physical contact to communicate care and affection, such as holding or patting the person's hand or hugging them. This provides reassurance to the person with dementia.
- Speaking clearly and calmly helps the person with dementia to follow the conversation.
- Speaking at a slightly slower pace and allowing time between sentences provides time for the person to process the information and to respond.
- Avoid speaking sharply or using a loud voice, as this may upset the person.
- Short and simple sentences increase comprehension.
- Avoid talking about someone with dementia as if that person were not present or talk in a way that would be appropriate to a young child. It is critical that respect and patience characterize communication.
- Validating the emotions behind the words and/or facial expression of the person with dementia is critical. The ability of the caregiver to focus on feelings helps them to discover the deeper meaning behind the verbal communication. For instance, the person with dementia looks angry. The caregiver says, "You are looking upset. Is there anything I can do to make you feel better?" Even if the caregiver has misinterpreted the feeling, he or she is still communicating a desire to understand and to improve the situation.
- Maintaining eye contact is important. The facial expressions and body language of the caregiver as well as the person with dementia communicate vital

information. The cared for person is able to "see" that the caregiver is interested and the caregiver is able to "see" the emotions of that person.

- Always gain the attention of the person with AD before talking. Calling the individual by name is essential.
- As noted before, avoid reasoning and arguing with someone who has dementia. It is better to accept and agree rather than to argue. No one wins an argument with a person who has dementia and arguing leads to agitation and sometimes to aggression.
- A positive approach many times elicits a positive response. Instead of saying, "Don't do that," say, "Let's try this."
- A quiet and calm atmosphere is important because those with dementia have diminished capacity to deal with a stressful and noisy environment. The stress of the environment has a contagious impact that often leads to agitation in the person.
- Always approach the person with dementia from the front. Peripheral vision deteriorates as the disease progresses. A person who is approached from the side might strike out because he or she doesn't see the approaching caregiver and becomes frightened by their sudden appearance.
- As repeatedly emphasized here, "park" negative emotions outside of the interaction with the person with AD who retains because they retain the ability to "sense" feeling states in others but lack the ability to validate the meaning of those feeling states. For instance, the person might sense tension and negativity in a caregiver but is unable to ask, "Did I do something to make you angry with me?" Instead the person may react to the perceived negativity in the caregiver as if he or she caused the caregiver to be tense. The person with dementia maintains skills to interpret feelings inherent in posture and voice tone but not to validate her or his conclusions. Inappropriate interpretations of nonverbal messages can lead to hurt or angry feelings. If the caregiver is feeling stressed and is unable to control his or her reactions, he or she should say, "I am tired; please don't think I am angry with you."
- Brutal honesty with the person who has dementia is unkind. If the person is looking for a deceased spouse who has been deceased for a long time, it is better to say, "He is not here right now" than to respond, "He is deceased and has been for ten years." Such a heartless response causes the person with dementia to grieve and to experience guilt because she has not remembered her husband's death or funeral. Because she will not remember the answer and will repeat the question about her husband's whereabouts, she will grieve every time a caregiver shares this truth with her.
- Avoid the question "don't you remember?" Family members also need to be told that this is not a good question to ask. If the person with dementia remembered he or she would not ask the same questions repeatedly. This question "puts the person on the spot" leaving him or her feeling anxious. Additionally, close friends or family members who are not recognized might feel hurt and offended. Imagine the sadness felt by grandchildren when their grandmother no longer recognizes them. In place of "don't you remember?" it is more effective and kinder to supply the forgotten information to the person with dementia. For example,

a person is visited by her daughter who wants to surprise her mother by bringing her youngest granddaughter to the visit. The daughter can save her mother embarrassment and the granddaughter hurt feelings by avoiding the question, "Mom I have a surprise for you. Do you know who this is?" A more effective approach would be, "Mom I brought Sarah to see you today—she is your youngest grandchild and I know how happy you are to see her." The latter approach saves the grandmother from feelings of embarrassment because she knows she should recognize Sarah but does not. It also saves Sarah from feeling sad because her grandmother does not recognize her.

- Assist the person with activities of daily living by giving them instructions one step at a time, using gestures to illustrate each step. The old saying, "in one ear and out the other" is absolutely true for the person in the stage 6 of the FAST scale. Short-term memory is so deteriorated that he or she will only be able to follow one step directions.
- Write directions in *big simple words,* label drawers, make picture albums that show the adult children and adults with a corresponding picture of when they were children. As noted above, tapping into reading as a remaining skill opens up tremendous ways to help the person with dementia "seem" and "feel" more independent than is true in reality.
- Avoid quizzing—it does nothing to help a person with dementia to remember but it is a very effective strategy to increase agitation.
- Use story telling. Timeslips$_{\text{TM}}$ is a nationally recognized program that helps caregivers to use the power of stories, generated by the person with dementia, to improve communication and mood.

Specific Challenging Situations

There are specific communication challenges that call for more guidance. For instance, when a caregiver is confronted with escalating behaviors in the cared for person, the following techniques are designed to calm down the person:

1. Back off from the individual. Avoid being close enough to be hit, bitten or physically harmed.
2. Control personal anxiety by taking several slow deep breaths before speaking—quietly count to ten.
3. Control nonverbal communication including facial expression and muscle tone.
4. Quietly validate the behaviors that are being displayed, "I can see you are upset. Can we have a cup of tea together while you tell me what is bothering you? "
5. As always, try to determine whether the individual is in pain or has another physical need that might be driving agitated behaviors.

Another situation that frequently occurs in caregiving situations is that the person with dementia accuses the caregiver of stealing an object such as a purse. In situations of this nature, try the following:

1. First, avoid "putting up a defense". Defending against a false accusation has the same effect as throwing gasoline on a fire—the person will become increasingly agitated, more difficult to console and could become aggressive.
2. As noted above, control personal anxiety by taking several slow deep breaths before speaking—quietly count to ten or say a prayer.
3. As noted above, control nonverbal communication including facial expression and muscle tone.
4. Say, "Oh Mom is your purse missing again? I know how upset you must be (the caregiver is validating her or his mother's feelings). When did you see it last? What was in it? Was there anything really important in that purse? Let me help you find it (offering to help). The caregiver begins to look and while looking for the purse starts to talk about other topics (using redirection). If the caregiver can maintain this banter for 5 min or more, the person with AD will most likely forget the accusation. This does not guarantee that Mrs. Smith will never accuse that family member or a paid caregiver again, but if this approach works, it will work every time.

An individual with Alzheimer's could accuse his/her spouse of cheating, it could stem from fear that he/she will be abandoned by the spouse or other family members. This fear and resulting accusations can also arise because the person with Alzheimer's is beginning to have difficulty recognizing people and may interpret the actions of an acquaintance, who is no longer recognized, as the actions of someone who is "hitting on her/his spouse".

According to Geri R Hall, Ph.D., ARNP, GCNS, FAAN Clinical Nurse Specialist Banner Alzheimer's Institute in a CARING.COM blog, she suggested the following technique when a spouse is dealing with accusations of cheating:

a. *Acknowledge the accuser's distress,* "You think I was cheating. Oh John, I am so sorry you feel that way. I will do anything you suggest to correct it!"
b. *Apologize again:* "I am so very sorry you think that I was cheating on you. I love you and have no intention of leaving you. I will try to never upset you again."
c. *Agree:* "If I thought that you were cheating I would be as upset as you are."
d. *Play dumb* and promise to fix it…even cry a little. "Oh John, I don't know how this happened. I will make sure it never happens again. I'm here with you and have no plans to go anywhere."
e. *Redirect and ask:* Why don't we get some ice cream? What flavor would you like? (Ice cream can sometimes be a great substitute for mood-controlling medications!)

Summary

Effective communication is vital to all that we do, and this is certainly true for those who care for someone with AD or other dementia. Caregivers are faced with communication challenges that demand responses and approaches that are unnecessary with those who are cognitively intact. Although persons with dementia continue to be active communicators, their skills in speaking, listening, and interpreting messages are all radically impacted by their disease. This means that the caregiver needs to try different approaches. This chapter provides many alternate ways of communicating with someone with AD or other dementia. The strategies laid out in this chapter are loving, respectful, and designed to support the continuation of a meaningful relationship between caregivers and those for whom they provide care.

References

Algase, D. et al. (1996). Need-driven dementia-compromised behavior: an alternative view of disruptive behavior. *American Journal of Alzheimer's Disease and Other Dementias, 11*(6), 10–19.

Baddeley, A. (1995). The psychology of memory. In A. D. Baddeley, B. A. Wilson, & F. N. Watts (Eds.), *Handbook of memory disorders* (pp. 3–26). New York: Wiley.

Bourgeois, M. (1990). Enhancing conversation skills in Alzheimer's disease using a prosthetic memory aid. *Journal of Applied Behavior Analysis, 23,* 29–42.

Bourgeois, M. (1992). Evaluating memory wallets in conversations with patients with dementia. *Journal of Speech and Hearing Research, 35,* 1344–1357.

Bourgeois, M. (1993). Effects of memory aids on the dyadic conversations of individuals with dementia. *Journal of Applied Behavior Analysis*, *26*, 77–87.

Bourgeois, M. (1997). *My book of memories: A workbook to aid individuals with impairments of memory*. Gaylord: Northern Speech Services.

Bourgeois, M. (2001). Matching activity modifications to the progression of functional changes. In E. Eisner (Ed.), *"Can do" communication and activity for adults with alzheimer's disease: strength-based assessment and activities* (pp. 101–107). Austin: Pro-Ed.

Bourgeois, M. (2007). *Memory books and other graphic cuing systems*. New York: Health Professions Press. (www.amazon.com or www.healthpropress.com).

Bourgeois, M. & Hickey, E. (2009). *Dementia: From diagnosis to management—A functional approach*. New York: Taylor & Francis.

Bourgeois, M. & Irvine, B. (1999). *Working with Dementia: Communication Tools for Professional Caregivers*. CD-ROM and Videotape In-service training programs available from ORCAS, Oregon Center for Applied Science, Inc., Eugene, OR. Award of Distinction from The Communicator and Finalist 2000, Telly Awards.

Bourgeois, M., & Mason, L. A. (1996). Memory wallet intervention in an adult day care setting. *Behavioral Interventions: Theory and Practice in Residential and Community-based Clinical Programs, 11,* 3–18.

Bourgeois, M., Burgio, L., Schulz, R., Beach, S., & Palmer, B. (1997). Modifying repetitive verbalization of community dwelling patients with AD. *The Gerontologist, 37,* 30–39.

Bourgeois, M., Schulz, R., Burgio, L., & Beach, S. (2002). Skills training for spouses of patients with Alzheimer's disease: Outcomes of an intervention study. *Journal of Clinical Geropsychology, 8,* 53–73.

Bourgeois, M., Camp, C., Rose, M., White, B., Malone, M., Carr, J., & Rovine, M. (2003). A comparison of training strategies to enhance use of external aids by persons with dementia. *Journal of Communication Disorders, 36,* 361–379.

Bourgeois, M., Dijkstra, K., Burgio, L., & Allen, R.S. (2004). Communication skills training for nursing aides of residents with dementia: The impact of measuring performance. *Clinical Gerontologist, 27,* 119–138.

Bourgeois, M., Dijkstra, K., & Hickey, E. (2005). Impact of communicative interaction on measuring quality of life in dementia. *Journal of Medical Speech Language Pathology, 13,* 37–50.

Brooker, D. (2004). What is person-centered care in dementia? *Reviews in Clinical Gerontology, 13,* 215–222.

Brush, J.A. & Camp, C.J. (1998). *A therapy technique for improving memory: Spaced retrieval.* Beachwood: Menorah Park.

Brush, J., Calkins, M., Bruce, C., & Sanford, J. (2011). *Environmental and communication assessment toolkit for dementia care.* Health Professions Press.

Burgio, L., Allen-Burge, R., Roth, D., Bourgeois, M., Dijkstra, K., Gerstle, J., Jackson, E., & Bankester, L. (2001). Come talk with me: Improving communication between nursing assistants and nursing home residents during care routines. *The Gerontologist, 41,* 449–460.

Camp, C. (1999). Memory interventions for normal and pathological older adults. In R. Schulz, G. Maddox, & M. P. Lawton (Eds.), *Annual review of gerontology and geriatrics: Focus on interventions research with older adults* (pp. 155–189). New York: Springer.

Carson, V. B. (2011). Responding to challenging behaviors in those with Alzheimer's: communication matters. *Caring Magazine, 30*(3), 26–31.

Carson, V. B., & Smarr, R. (2006). Alzheimer's disease: Increasingly the face of home care. *Caring Magazine, 25*(5), 6–10.

Carson, V. B. & Smarr, R. (2007). Becoming an Alzheimer's whisperer. *Home Healthcare Nurse, 25*(10), 628–636.

Dijkstra, K., Bourgeois, M., Youmans, G., & Hancock, A. (2006). Implications of an advice giving and teacher role on language produce in adults with dementia. *The Gerontologist, 46,* 357–366.

Hickey, E., & Bourgeois, M. (2000). Measuring health-related quality of life (HR-QOL) in nursing home residents with dementia. *Aphasiology, 14,* 669–679s.

Hoerster, L., Hickey, E., & Bourgeois, M. (2001). Effects of memory aids on conversations between nursing home residents with dementia and nursing assistants. *Neuropsychological Rehabilitation, 11,* 399–427

Irvine, B., & Bourgeois, M. (2004). *Professional dementia care: skills for success.* CD-ROM and Videotape In-service training programs available from ORCAS, Oregon Center for Applied Science, Inc., Eugene, OR.

Irvine, A. B., Bourgeois, M., & Ary, D. V. (2003). An interactive multi-media program to train professional caregivers. *Journal of Applied Gerontology, 22,* 269–288.

Irvine, A. B., Bourgeois, M. S., Billow, M., & Seeley, J. (2007). Web Training for CNAs to Prevent Resident Aggression. *Journal of the American Medical Directors Association,* in press.

Kitwood, T. (1997). *Dementia reconsidered: the person comes first.* Buckingham: Open University Press.

Palmer, C., Adams, S., Bourgeois, M., Durrant, J., & Rossi, M. (1999). Reduction in caregiver-identified problem behaviors in patients with Alzheimer Disease post hearing-aid fitting. *Journal of Speech, Language, Hearing Research, 42,* 312–328.

Rader, J. (1995). *Individualized dementia care: Creative, compassionate approaches.* New York: Springer.

Sabat, S. (2001). *The experience of Alzheimer's disease: Life through a tangled veil.* Oxford: Blackwell.

Squire, L. (1994). Declarative and nondeclarative memory: Multiple brain systems supporting learning and memory. In D. L. Schacter & E. Tulving (Eds.), *Memory systems* (pp. 203–232). Cambridge: MIT Press.

Thomas, W. H. (1996). *Life worth living: How someone you love can still enjoy life in a nursing home. The Eden Alternative in action.* Acton: Vanderwyk and Burnam.

Wicker, C. (2010). Unlocking the silent prison. *Parade Magazine, 11,* 34. http://parade.condenast.com/105537/christinewicker/21-unlocking-the-silent-prison/. Accessed 1 May 2014.

World Health Organization (2002). *International classification of functioning, disability and health (ICF).* Geneva: World Health Organization. (Chicago: Alzheimer's Association. 2006. 5 p.).

Chapter 9
Whispering Hope and Faith: Still, Small Voices for the Alzheimer's Journey

Profound memory loss is commonly referred to as "loss of self." However, it never means "loss of soul." People with Alzheimer's disease (AD) continue to experience spiritual needs, as do their caregivers. For many people, religious or spiritual experiences are part of their long-term memories. These memories are accessible to the person with AD until very late in the disease (Ewing 2005; Killick 2006). Ancient memories of a parent praying with a child or singing hymns to comfort may date back to infancy and so are preserved in the brain well after verbal communication has been destroyed along with short-term memory. Religious music continues to be one of the most powerful interventions for reaching those in the late stages of AD. Music is stored in the brain through a complex neural network that preserves it when other areas of the brain are destroyed through AD. This chapter discusses the many ways that spiritual needs can be met through music, literature, touch, prayer, candles, a menorah, a rosary or Bible, the smell of incense, wine, or fresh bread (Beuscher and Beck 2008).

AD alters every dimension of life including the spiritual. The individual's ability to communicate what is spiritually meaningful begins to diminish along with all of the other cognitive functions. However, this does not mean that caregivers should ignore the issue of spirituality and/or assume that the person with dementia is unable to respond to the spiritual. As a caregiver, it is easy to experience a sense of futility when working with a loved one or a patient in a facility. It is understandable that caregivers might conclude that it would be futile to even attempt to meet spiritual needs because the person under their care will not understand. Nothing could be further from the truth. Many individuals were exposed to religious beliefs, experiences, and practices beginning in infancy (Ennis and Kazer 2013). Perhaps family members said nighttime prayers during this time—a prayer as simple as:

> Now I lay me down to sleep, I pray the Lord my soul to keep. If I should die before I wake, I pray the Lord my soul to take.

Perhaps family members sang religious hymns to the child, or perhaps he or she was exposed to a religious rite such as baptism and later to first communion and confirmation or perhaps the child's parents dedicated him or her to God. Perhaps

V. Benner Carson et al., *Care Giving for Alzheimer's Disease*,
DOI 10.1007/978-1-4939-2407-3_9

the child experienced a bar mitzvah or learned prayers in Hebrew, Farsi, or another language. These early experiences may be imbedded deep within the brain in the individual's long-term memory. The person may not remember the name of their most recent clergy person or what faith tradition they belonged to, but these lapses do not guarantee that the person has forgotten who God is. Individuals with AD may still remember God, and will respond to prayer and hymns. When the opportunity to actively or passively participate in prayer and singing religious music is offered, many of those with AD will join in with these activities (Gataric et al. 2010).

A video clip, embedded within a larger video entitled *Memory Bridge* shows a vignette of Naomi Feil interacting with Gladys Wilson who has not spoken in 2 years. (www.memorybridge.org/Accessed 4 May 2014.)

In the video, Naomi, a teacher and consultant on the care of those with AD, moves very close to Gladys and begins to gently touch Gladys's cheeks—the way a mother would stroke the cheeks of a baby. Naomi draws upon Gladys's religious history and uses old church hymns to connect with Gladys. Initially, Naomi, while continuing to stroke Gladys's cheeks, sings the children's hymn, *Jesus Loves Me, this I know, for the Bible tells me so.* Although Gladys does not join in the singing, she opens her eyes and focuses intently on Naomi's eyes; Gladys also begins to clap her hands on the arms of her chair as well as on Naomi's arms! Although Gladys does not verbally respond, her eyes remain fixed on Naomi's face. Naomi then asks Gladys if she will join Naomi in another song. Naomi begins to sing *He's Got the Whole World in His Hands* and when she gets to the second verse, Gladys begins to sing along! It is an incredibly powerful moment when at the end of the song, Naomi asks Gladys if she feels safe and warm, with Jesus and Naomi—to which Gladys responds yes! This short video clip is a powerful example of how religious music can reach down into the soul of a person and bring forth a response even from someone like Gladys who appears nonverbal and uncommunicative.

The religious beliefs, practices, and expressions of faith that are cradled in long-term memory can be tapped into long after short-term memories are gone, even during the last stages of AD. Profound memory loss is commonly referred to as "loss of self." However, it never means "loss of soul." The soul remains and caregivers can get there from where they are (Mooney 2004; Snyder 2003).

What are the implications of this knowledge for the caregiver? Church, synagogue, temple, or mosque are the formal places where people go to pray and to honor God regardless of religious tradition. People look to their faith to be accepted, spiritually challenged, and fed, and to receive answers and direction during the most difficult times of life. How are these needs met when the caregiver can no longer access the religious facility? Should family members approach the pastor, rabbi, imam, or temple leader to request that an announcement be made to everyone, including those members of the congregation with special needs, that they are welcome? Should members of a congregation expect that their faith community be educated from the pulpit about the special needs of members? Should the person with AD or other dementia still be transported to church, synagogue, or temple and be allowed to participate in the religious service in whatever manner is possible for the individual? What if the person with dementia, exhibits inappropriate behaviors in the religious setting? How is it possible for the caregiver to respond to these behaviors and still focus on the religious service? These are all questions that caregivers as well as religious leaders must confront. Considering the potential onslaught of those with AD as the baby-boomer generation races into "old age," it would behoove spiritual leaders to inform the congregation that not only are they a welcoming community to those who are cognitively intact but also to those who are cognitively impaired. The last thing that should happen is that family caregivers are "cut off" from spiritual comfort because their spiritual community does not accept or understand the special needs of those with AD (Grollman 2004). But even if the religious community threw open their doors to welcome those with AD, the issues that surround dealing with disruptive behaviors still remain (Stolley et al. 1999).

Impact of Memory Loss

Some suggest that memory loss leads to chaos in the soul and a sense of abandonment because it is memory that allows us to hold onto God's presence in our lives. We recall answered prayer, we remember calling out to God in pain and sorrow, and we remember God's response. Memory affects how we approach the future—we remember God's past with us, and we trust that God will be with us in our future. So it would seem logical to conclude that the memory loss from dementia cuts the cords of continuity that are meant to sustain the soul until the last breath. But is this so? Young children and infants depend on loved ones for their sense of God's presence—why would this not be true for adults whose cognitive and functional abilities are that of a very young child? It seems that another responsibility of the caregiver is to continue to represent God's love and grace to the person being cared for. It does not matter that the "idea" of God is no longer a possibility for the person with AD. The "reality" of love, concern, and care are pathways to God with or without cognitive awareness. Both caregivers and communities of faith have the privilege and shared responsibility to help those with dementia to find their way home to God in the midst of the chaos caused by the disease (Trevitt and MacKinlay 2004).

How can these challenges be met? What can caregivers do to help repair the spiritual connections and restore the way home to God? Actively using religious rituals and practices, talking about the beliefs of the person's faith tradition, praying with the person, and reading from religious texts such as the Bible or the Torah—all of these activities are activities that parents engage in with young children. While the child cannot grasp the idea of a God, he or she can certainly sense the feelings and sense of comfort that parents have when they talk about God. These conversations generally convey love, connectedness, and acceptance—and lead to feelings of safety and security—exactly what parents want to convey, and exactly what caregivers to those with AD must do also.

In order to restore those spiritual connections, caregivers need to be flexible about their definition of spiritual needs. First, there is a need to explore the "meaning-making" history of the person with AD. When families are instructed about this, they are usually able to identify the activities and beliefs that were part of their loved one's spiritual history. For instance, they will know whether their loved one regularly attended religious services; they will know that their father found spiritual solace from nature, or perhaps their mother connected with the divine through music. Every person possesses a "meaning-making history" Higgins (2005). By exploring that history, caregivers can arrive at activities that are spiritually significant for the person with AD. For instance, adult children might remember that their mother said the rosary multiple times every day but that she never used rosary beads; instead, she counted the beads of the rosary on her fingers. Perhaps for this person, listening to the recitation of the rosary on television, radio, or a compact disc (CD) would be a powerful spiritual intervention drawing on old and very powerful memories.

Perhaps the family may recall that when mom was in church, dad was usually involved in something outside—that he thrived on nature and found spiritual solace in the outdoors. This knowledge might suggest several "spiritual interventions" including regular walks outside, watching videos that featured the beauty of nature, or activities such as fishing and sightseeing. Meeting spiritual needs must be very flexible.

Sometimes the use of visual aids such as a menorah, candles, the Bible, a rosary, or figurines of praying hands may stimulate the recall of retained memories linked to the soul. Beyond visual cues, there are other sensual experiences such as the smell of incense; fresh bread or even wine might unlock powerfully meaning-making memories. Sounds such as recordings of church bells, the shofar horn, organ music, the words of scripture, or liturgy are all capable of eliciting spiritual memories and emotions. Religious music is probably the strongest bridge to the person's spirit.

The FAST Scale and Religious Needs

Let us look at meeting spiritual needs in light of the functional assessment staging tool (FAST) scale (Carr et al. 2011). In stage 4, when the person is considered "a great Fooler," those with AD can still draw on long-term memory and act in an appropriate manner during religious services. Memory loss is not obvious to others unless they live with the person. Faith symbols such as a Bible, cross, menorah, and/or religious pictures are still meaningful to the individual. He or she can actively participate in religious services, and family caregivers should be encouraged to not only attend services themselves but to also bring their cognitively impaired loved ones with them. The person with AD can also participate in activities within the faith community if they are highly structured, such as distributing hymnals to attendees at the beginning of the service and bulletins at the end of the service. Religious hymns are important "connectors" not only to the present religious service but also to old and possibly very distant but well-engrained emotional memories. Reading and reciting scripture and prayers draw on old memories and serve the same purpose as for those with or without AD. Prayer continues to be important for the person in stage 4 along with reminiscence about holidays, weddings, and other ceremonies with structured rituals that are stored in long-term memory. The presence of Bibles, religious pictures, and hymnals are powerful symbols that allow the person to draw on old memories, while still receiving spiritual benefit from the service. In this stage, however, the person might also be experiencing grief over his or her loss of memory or ability to function independently. The person is still capable of insight, and insight can prove to be very painful, as they may be aware of what is being lost. Beyond stage 4, the person will forget the losses—that parents, a spouse, or other significant individuals are gone. A stage 4 person may seek the comfort that these people would have offered, and when they are not available, may suffer. Caregivers can validate the loss and suffering but cannot take it away.

Religious professionals can be a source of great solace to those with AD by making sure that all those in their religious congregations regularly receive communion or other sacraments (Thompkins and Sorrell 2008). This is especially true for the person with AD who is homebound—regular reception of communion maintains a link with the faith community. If a church community has a faith-community nurse on staff, this person can make home visits and bring a spiritual message from the church community including communion. Additionally, the nurse will be able to assist the caregiver with managing the care of their loved one (for example, by mobilizing respite support).

In stage 5, when the person with AD is functioning at the level of a 5–7-year-old, he or she will likely display behaviors that are not appropriate in a formal religious setting. First of all, without supervision, the person at this stage might dress inappropriately, i.e., putting on underpants over dress slacks or wearing a low-cut dress that exposes part of the woman's breast. The potential for these behaviors places a responsibility on family members or other caregivers to supervise dressing. Additionally, the person with dementia might yell out inappropriate comments during the religious service—this type of behavior indicates that the family should sit with their loved one in the "cry" room that many religious facilities provide for very young children—close to the exit—so that they can make a quick getaway if needed during the service.

In stage 6, when the person is functioning at the level of a toddler and displaying behaviors such as extreme restlessness, mood swings, and confusion, it may be difficult if not impossible for the person with AD to participate in services. If members of the faith community are trained on how to relate to those with AD, then they could spend time with the person during the service in a location separate from the sanctuary. This would allow the primary caregiver to fully participate in the service and to receive spiritual consolation and support.

During stage 7, the last stage, the person with AD is functioning at a cognitive and functional level of about 18 months to newborn. Most of the time, it is not possible for the person with dementia to attend a formal religious service. In this stage, he or she is losing the ability to walk and is totally incontinent. This is the time when family members, good friends, and members from the church could provide "sitter" services at home, so that the primary caregiver is still able to receive the spiritual support at church that is so necessary to be a good caregiver.

Spiritual Needs of the Caregiver

Being a family caregiver for a loved one at home is difficult. The commitment to faithfully provide care 24 h a day over many years represents the embodiment of love. However, even the most loving caregivers need respite. They need to replenish their own spiritual, emotional, and physical resources. The love of friends and family directed to the caregiver reminds the person that he or she is not alone and that there are others who can be relied on to help. Caring and spiritual responses from family and friends are specific offers of support. "Give me a call if you need any help" is not a valid offer and does little to make the caregiver feel "cared for." Rather the caregiver who is usually stressed and overwhelmed needs "helpers" who offer concrete assistance. For example, "Mom, I will come over every Wednesday night and stay with dad so you can continue to go to choir practice." Or, "Dad, I know you like to play cards with your buddies a couple of times a month—let's set up a schedule so that I can plan to be here with mom while you are out."

Even with support, not every caregiver is able to provide this level of care. Those who recognize that the amount of care required is beyond his or her physical and emotional capabilities should never be judged for a decision to place a loved one in a nursing home or other facility. Until family and friends have walked in that caregiver's shoes and experienced the relentless demands of caregiving, all judgment should be suspended. The only appropriate responses from family and friends are offers of help and loving support. The recognition that a caregiver has reached his/her limit is not a sign of weakness or lack of love but a realistic appraisal of the situation.

Additional strategies to help caregivers through this process include providing a journal and encouraging the caregiver to keep a record of the caregiving experience and the spiritual understanding involved in those experiences. Writing can be a remarkable anecdote to stress. Writing allows the person to express frustration, anger, joy, and feelings of success, abandonment, and other reactions that accompany caregiving. Visits from members of the person's place of worship, including the faith-community nurse, are important times that let the caregiver know that they have not been forgotten. Any connection coming from the caregiver's faith community serves as a powerful link that provides comfort, and decreases the sense of isolation that is so often part of the caregiver's experience.

Summary

Spirituality continues to play an important role in the life of the person with AD as well as in the life of his/her caregiver (Jolley et al. 2010). The recognition of the importance of spirituality needs to translate into thoughtful strategies that revive and support a continued spiritual connection—not only for the person with dementia but also for the caregiver. Regardless of the unrelenting ravages of disease, spiritual needs continue, and the extent to which these needs are met can powerfully impact the physical, emotional, and spiritual health of the caregiver, while at the same time lovingly ushering "home" the person with AD.

References

Beuscher, L., & Beck, C. (2008). A literature review of spirituality in coping with early-stage Alzheimer's disease. *Journal of Clinical Nursing, 17*(5a), 88–97.

Carr, T. J., Hicks-Moore, S., & Montgomery, P. (2011). What's so big about the 'little things': A phenomenological inquiry into the meaning of spiritual care in dementia. *Dementia, 10*(3), 399–414.

Ennis, E. M., & Kazer, M. W. (2013). The role of spiritual nursing interventions on improved outcomes in older adults with dementia. *Holistic Nursing Practice, 27*(2), 106–113.

Ewing, W. A. (2005). Land of forgetfulness: Dementia care as spiritual formation. *Journal of Gerontological Social Work, 45*(3), 301–311.

Gataric, G., Kinsel, B., Currie, B. G., et al. (2010). Reflections on the under-researched topic of grief in persons with dementia: A report from a symposium on grief and dementia. *American Journal of Hospice & Palliative Medicine, 27*(8), 567–574.

Grollman, E. A. (2004). Spirituality and Alzheimer's disease. In *Living with grief: Alzheimer's disease* (pp. 213–223). Washington: Hospice Foundation of America.

Higgins, P. (2005). Bringing spiritual light into dementia care. *Journal of Dementia Care, 13*(2), 10–11.

Jolley, D., Benbow, S. M., Grizzell, M., et al. (2010) Spirituality and faith in dementia. *Dementia, 9*(3), 311–325.

Killick, J. (2006). Helping the flame to stay bright: Celebrating the spiritual in dementia. *Journal of Religion, Spirituality & Aging, 18*(2–3), 73–78.

Mooney, S. F. (2004). A ministry of memory: Spiritual care for the older adult with dementia. *Care Management Journals, 5*(3), 183–187.

Snyder, L. (2003). Satisfaction and challenges in spiritual faith and practice for persons with dementia. *Dementia, 2*(3), 299–313.

Stolley, J. M., Buckwalter, K. C., Koenig, H. G. (1999). Prayer and religious coping for caregivers of persons with Alzheimer's disease and related disorders. *American Journal of Alzheimer's Disease, 14*(3), 181–191.

Thompkins, C., & Sorrell, J. M. (2008). Older adults with Alzheimer's disease in a faith community. *Journal of Psychosocial Nursing and Mental Health Services, 46*(1), 22–25.

Trevitt, C., & MacKinlay, E. (2004). "Just because I can't remember…": Religiousness in older people with dementia. *Journal of Religious Gerontology, 16*(3–4), 109–121.

Chapter 10
What Else…?

This chapter addresses the question, "What else is there to know about managing dementia?" There are many other issues that could be addressed, probably more than this book has space to cover. However, there are some behaviors that are almost universally present that lead to management challenges for caregivers. Issues such as the importance of keeping good records; removing the car keys; taking over money management; responding to sundowning behavior; explaining to grandchildren why a grandparent no longer recognizes them; handling holidays and vacations; and last, making the decision to place a loved one in either an assisted living facility or a long-term skilled facility. Each of these behaviors is addressed from the perspective of the theory of retrogenesis (see Chap. 1).

Importance of Keeping Good Records

It is valuable when speaking to a physician or another care provider such as a home health-care nurse to be able to (1) quantify the frequency of challenging behaviors, (2) report the appearance of a new challenging behavior or a change in behavior, (3) indicate the time the behavior occurs, and (4) provide details regarding what else was taking place at the time of the challenging or changed behavior. Likewise, if the caregiver is instituting a new intervention to manage a challenging behavior (such as starting a new medication), quantifying the response is invaluable. Keeping records provides important information to professional caregivers and may be vital in guiding their decisions regarding the effectiveness of a change in the treatment plan. The following table serves as an example of the type of record keeping that is useful to caregivers.

Using Behavior Logs

The Value of Tracking the "Who, What, and When" Behind Challenging Behaviors.

V. Benner Carson et al., *Care Giving for Alzheimer's Disease*,
DOI 10.1007/978-1-4939-2407-3_10

USING BEHAVIOR LOGS

The Value of Tracking the "Who, What and When" Behind Challenging Behaviors

Dementia Behavior Analysis					
Time	Activity	Behavior	People Around	Medication	Comment
7:00					
7:00					
7:30					
8:00					
8:30					
9:00					
9:30					
10:00					
10:30					
11:00					
11:30					
12 Noon					
12:30					
1:00					
1:30					
2:00					
2:30					
3:00					
3:30					
4:00					
4:30					
5:00					
5:30					
6:00					
6:30					
7:00					
7:30					
8:00					
8:30					
9:00					
9:30					
10:00					
10:30					
11:00					
11:30					
12 Midnight					
12:30					
1:00					
1:30					
2:00					
2:30					
3:00					
3:30					
4:00					
4:30					
5:00					
5:30					
6:00					
6:30					

An example will serve to demonstrate how essential it is to keep good records.

> A daughter was caring for her mother in the daughter's home. The daughter complained that every day around 3PM her mother "fell apart" and would become highly agitated, would cry and might even strike out at her daughter. The home care nurse encouraged the daughter to complete the table above in order to see if there were patterns missed by the daughter that could shed light on this challenging behavior. After completing the chart for four days, the pattern became crystal clear to both the daughter and the home care nurse. The three grandchildren began arriving home at 3PM—the noise level and confusion in the home markedly increased from 3PM until about 4:30 pm. The daughter, who was busy listening to her childrens' accounts of their days and inquiring about homework, was oblivious to

the change in the atmosphere in the home. The home care nurse suggested a number of interventions. The first was to move her mother into a room where she would be protected from the noise of the homecoming children and where she could watch DVD's of "old" movies and/or listen to music that she loved. The daughter would tell her mother that this was "mom's special time to do what she enjoyed," and then brew a cup of tea for her mother and provide a snack to increase her blood sugar. After the children had settled down and the environment became more peaceful, she would send each of the children into grand mom's room to visit with her. It was not until the daughter began to keep records that she saw the connection between her mother's agitation and the time of the children's homecoming.

Taking Away the Car Keys

For most of us, being able to drive is a sign of maturity and becoming an adult. Driving allows us a great degree of independence to "to come and go" as we choose; to handle the routine activities that allow us to function—going to the bank, doing grocery shopping, making a trip to the hardware store; to see friends, go to church, synagogue, or mosque whenever we want; to go to the barber or hairdresser; and, when necessary, to see the doctor. When it becomes clear that a loved one is no longer safe to drive, all of these "trips" fall on the shoulders of the caregiver(s). Many caregivers at this time may still be involved in the lives of young children, and/or the caregiver may be working outside of the home. Individuals with dementia in stage 4 on the functional assessment staging tool (FAST) scale, sometimes called the Great Foolers, may independently modify their driving by limiting nighttime outings and trips to unfamiliar locations. However, as the disease progresses, the person with dementia loses the ability to accurately evaluate her/his thinking and his/her response time increases—so that driving skill deteriorates without self-awareness. The challenge is to preserve the individual's sense of independence as long as possible while at the same time protecting that person's safety and the safety of others.

Caregivers sometimes allow driving to continue when it is clear that it is no longer safe. At one extreme, some caregivers may not want to "hurt the feelings" of the person and so allow driving to continue long past the time when it is no longer safe. At the other extreme, there are caregivers who want to "take away the keys" at the earliest sign of a memory deficit. Some caregivers may overreact when a loved one with dementia fails to come to a complete stop at a stop sign. Still other caregivers will need the support of family, friends, and professionals before making the decision to take away the keys to the car.

Once a person has been diagnosed with dementia, it is important for family and/or friends to be aware of changes in that person's driving skills. The Hartford publishes a helpful resource guide entitled *Warning Signs for Drivers with Dementia* that provides family and friends with concrete behaviors that aide in making the decision to remove the keys (See appendix at the end of chapter). Many states offer driving evaluations through the Motor Vehicle Administration, and the physician can support the family by suggesting that the patient be tested to ensure driving safety and competence.

Regardless of whether the physician suggests a driving assessment or the family initiates the assessment, it is an extremely difficult issue to address. Not only is it important that the patient not drive if her/his driving is deemed unsafe but providing transportation is also a huge responsibility for the family. Checking out resources in a community such as "elder ride" services and public transportation may be helpful. The more alternatives there are to driving, the easier the adjustment will be. It is important for the person to continue to get out for essentials like doctor's appointments, but also for social visits and enrichment. Feeling housebound can quickly lead to depression.

When driving becomes a safety issue, it is probably time to evaluate the living arrangements of the person with dementia. If the person lives in a remote area with few transportation options, it may be necessary to consider relocating to an area with more options, or to investigate senior living. Some of the transportation options include:

- Public transportation. If available, there are usually reduced rates for older adults.
- Ride sharing is another possibility in some communities.
- Community shuttles/senior transit. Many communities may have shuttle services available, especially for medical appointments. Some medical facilities, such as those for veterans, also have transportation options for medical appointments. Local churches may also offer transportation assistance.
- Taxis or private drivers. Taxis may be a good option for quick trips without a lot of prior scheduling.
- Walking/cycling. If health permits, walking or cycling is a great way not only to get around but also to obtain some exercise. Regular physical activity lowers the person's risk for a variety of conditions, including Alzheimer's and dementia, heart disease, diabetes, colon cancer, high blood pressure, and obesity.
- Motorized wheelchairs can be a good way to get around if the person lives in an area with easily accessible stores and well-paved streets.

Sundowning

The term "sundowning" refers to a state of confusion that occurs at the end of the day and into the night. If the person has her/his days and nights confused and sleeps all day, sundowning might occur at midnight or in the early morning hours. The behaviors that accompany sundowning may include confusion, pacing, crying, anxiety, aggression, or ignoring directions—basically, an emotional "meltdown." The exact cause of sundowning is unknown. Applying the theory of retrogenesis suggests a way to respond to sundowning behavior. When those with dementia experience sundowning, they are usually in stage 6 on the FAST scale or in the moderately severe stage of dementia. The person is functioning at a toddler level.

Consider the following scenario. A mother takes her three-year-old toddler to the pediatrician for a well-baby visit. The visit is scheduled at 2 p.m.—a time when the toddler is usually taking a nap. Following the appointment, the child is fussy, difficult to console, crying off and on again—and generally unpleasant the rest of the evening. The mother resolves that the next day she will be sure that the child will be back on schedule! How does this correlate to sundowning in the older person? Generally, those with Alzheimer's are at their best early in the morning and deteriorate as the day progresses (unless they sleep all day and awaken late in the day—in this case, sundowning will probably occur around midnight). Sundowning generally occurs late in the day—perhaps when a caregiver is making dinner and/or tending to children returning from school. The first episode of sundowning often catches the caregiver off guard, and great efforts are made to comfort the loved one with dementia. When this pattern occurs, the following is often helpful. About an hour before the sundowning occurs, the caregiver should stop what he/she is doing, sit with the person with dementia, provide a snack—a drink and something to eat—and engage the person in a quiet but pleasant activity such as listening to music from that person's era or perhaps look at picture albums together or watch an old black-and-white movie. By doing this, the caregiver is helping to restore the person with dementia so that he/she does not experience sundowning. If the person with dementia sleeps all day and awakens around 5–6 p.m., then the same strategies might be employed, but rather than intervening in the afternoon, the soothing interventions should occur around 11 p.m.

In addition to these strategies, there are videos that are made specifically to calm and engage the person with Alzheimer's disease (AD; see www.alzheimersvideo.com/). Another valuable website features 13 respite videos that provide caregivers with a way to improve his/her quality of life in this regard (see http://www.best-alzheimers-products.com). Using the respite videos provides a reprieve from the challenges of caregiving. All are designed to hold the attention of those with dementia through music, light movement, and reminiscence. The people on the videos become real friends to the person with dementia. As they have a conversation, or sing together, or remember times in the past with the people on the videos, the caregiver is able to experience a break from the caregiving role, time to prepare a meal, to write a letter, or to read a book. These videos have been successfully used in assisted living and skilled nursing facilities where the "change of shift times" with staff leaving a unit while others are coming into work can be quite chaotic. This is a time when anxiolytics and antipsychotic medications are often used to quiet residents who become disturbed with any change in routine. The videos can serve as an alternative to medication.

These tapes are wonderful additions to an overall care strategy. Each provides physical exercise in the form of light movement, socialization, reminiscence, and sensory and cognitive stimulation. All are ideal aids for caring for a person with dementia in the home, and they also work well in a group situation, in nursing homes, and Alzheimer's day care facilities.

Vacations

Caregivers of those with dementia when planning a vacation need to be aware that a vacation may not be the restful and relaxing get away that the caregiver had in mind (Carson 2012). Those with dementia do not always experience a vacation as a relaxing or fun time. In fact, traveling to unfamiliar locations might precipitate a catastrophic reaction in the person with Alzheimer's or other dementia. A caregiver might be lulled into a false sense of security based on the person's behavior at home and wrongfully conclude that the person with dementia will function just as well on a vacation. People with dementia do best with "sameness." Responding to unfamiliar places, persons, situations, and a different schedule every day can be a recipe for a catastrophic reaction. A vacation disrupts all of that "sameness" and familiarity, and plunges the person into situations that are not only unfamiliar but also potentially frightening.

An individual's reaction to a vacation is impacted by the degree of decline that has occurred due to dementia. A person in stage 4 ("the Great Fooler") is probably able to handle with relative ease the changes that accompany a vacation. However, for someone functioning at the level of a toddler, i.e., 4 years old deteriorating to 2 years old, a vacation could be stressful for everyone involved. In fact, the answer for a caregiver who desires to take a vacation might be to consider placing the person with AD in respite care—which is available either through certain home health-care agencies, assisted living facilities, or by bringing into the home a familiar individual, a friend, or a family member, to provide care so that the person with AD remains at home following familiar routines. Such an arrangement allows for someone to stay in the home while the caregiver is away.

Sometimes, the primary caregiver decides that including the loved one with Alzheimer's on a vacation is worth the effort. Planning and preparation, however, is vital in order to lessen the stress and make the vacation safer. It may help to take a short "test" trip to see how the person with dementia reacts to traveling. Planning the vacation destination for a location that is familiar to the person with AD and avoiding places that are overcrowded are also good strategies to consider. If the vacation includes visiting relatives and/or friends who are not aware of the changes that have occurred with dementia, it makes sense to forewarn those individuals about what to expect. For instance, informing them that "Sam still likes to use his hands to build things, but these days, he focuses on manipulating and building with Legos" gives them a heads-up regarding Sam's condition. If the person with dementia is a wanderer, it is essential that he/she is enrolled in the Alzheimer's Association's Safe Return program and that he/she wears an identification bracelet before leaving on a vacation. If the vacation involves flying and/or staying in a hotel, it is also important for the caregiver to inform the airline and hotel staff that she/he is traveling with a memory-impaired individual.

The caregiver should also plan to address the special needs of the person with AD. For instance, what are the bathroom needs of the individual—is the person incontinent, and if so, how will this be handled on a trip? Packing ample incontinent products, wipes, and changes of clothing that are easily accessible when needed is an important consideration. Beverly Bigtree Murphy (http://bigtreemurphy.com. Accessed 11 July 2012) created a website that includes practical strategies for dealing with incontinence during automobile travel. If traveling by car, Bigtree Murphy suggests to the caregiver look for service stations along main highways. As mentioned in Chap. 6, these facilities usually have a single-occupancy bathroom that not only provide privacy but will also have a sink that allows for cleanup from a bowel movement. This allows greater privacy for both the caregiver and the recipient of care. Major rest stops are usually equipped with multi-stall units with a sink outside of where the toilets are located so that there is no privacy for cleaning an incontinent loved one.

Packing should focus on changes of clothing and bringing enough prescribed as well as "over-the-counter" medications to deal with the unexpected physical issues that might arise (i.e., diarrhea, constipation, headache, upset stomach, etc.). Additionally, the planning for a trip needs to take into account activities that the person with AD can participate in while traveling. These include, for instance, loading an iPod with the person's favorite music, or bringing along a deck of cards to play with or, for some individuals, to just hold and manipulate (which may be soothing).

If possible there should be at least one additional person in the car who can focus on keeping the individual with AD calm, occupied, and safely secured with a seat belt. The itinerary for the trip needs to include regular rest stops. If the caregiver is traveling alone with the person and the individual with AD becomes agitated, it is important for the caregiver to stop the car and attend to the person's needs. Driving while trying to calm an agitated individual is guaranteed to result in disaster.

Air Travel

Traveling through airports requires plenty of focus and attention on the caregiver's part. The level of activity in an airport can be distracting, overwhelming, or difficult to understand for someone with dementia. It is especially important to avoid scheduling flights that require tight connections. Also helpful is to use the airport escort services to assist the person with dementia and caregiver to get from place to place. Even if walking is not difficult, the caregiver should request a wheelchair so that an airport employee is assigned to assist in getting the person with dementia and his/her caregiver to the correct location.

Prior to air travel, it is important to inform the airline and airport medical service department ahead of time that travel is planned with a memory-impaired individual with special needs. Once in the airport, it is also helpful to tell airport employees, screeners, and in-flight crew members. It is also important for the caregiver to carry the following documents:

- Doctors' names and contact information
- A list of current medications and dosages

- Phone numbers and addresses of the local police and fire departments, hospitals, and poison control (at the place they are traveling to)
- A list of food or drug allergies
- Copies of legal papers (living will, advanced directives, power of attorney, etc.)
- Names and contact information of friends and family members to call in case of an emergency
- Insurance information (policy number, member name)

Explaining to Grandchildren

Young children, when faced with a grandparent who is changing because of Alzheimer's, will invariably ask "what's wrong with grandma or grandpa?" Children need explanations. They are attuned to the emotional climate that they live in and are sensitive to signs that their parents are worried or upset. Without an explanation of what is going on, children will assume that they are responsible for the tension in the home or have done something wrong to make mommy and daddy upset. The first step is to explain to children that Alzheimer's is a sickness that changes the way grandma or grandpa acts and makes her/him forget things. The details that are provided to the child depend on the child's age. For instance, it makes sense to tell an older child that Alzheimer's is a sickness that changes grandma's brain on the inside (a young child might not know what a "brain" is and might be frightened or confused by complex explanations). Grandma may look the same on the outside but her brain is slowly changing from the inside. These changes lead to forgetting lots of important things, including people, even very important people like grandchildren that they love (Carson 2012).

Children will need reminders when the grandparent displays challenging behaviors that grandma or grandpa is really ill. If children witness catastrophic reactions where grandma is crying and/or screaming, they may again assume that they have done something wrong to make grandma behave this way. It is not unusual for children to carry an enormous burden of undeserved guilt for behaviors for which they bear no responsibility. They need reassurance, including repeated explanations that grandma is sick, that she loves them, and that they remain an important part of her life.

Sometimes, children are the targets of accusations of stealing or other wrongdoing because grandma cannot remember where she placed her purse or keys. Out of frustration, she blames any and all those around her of stealing. These accusations are difficult for adult family members to accept and even more difficult for children to handle. Very young children require comfort and consolation and assurance that they are not to blame. School-age and older children can be taught how to respond to grandma in a helpful manner. The following technique requires practice, but can help to defuse accusations and communicate love to grandma. The technique involves validating that grandma is upset because she lost her purse, slowly changing the subject (using the knowledge that grandma only has about 5 min of short-term memory), and relying on the power of distraction to move away from the false accusation. Rather than verbally denying wrongdoing, the child might say, "Oh, Grandma, I am so sorry that your purse is missing…please don't worry—I will help you find it. When I lost my favorite doll, I was so upset and you helped me find her. I will help you find your purse. Let's look for it together. I am sure it is somewhere—we just have to look. Where have you looked for it? Did you search in the closet? Under the bed? I remember that we found my doll underneath the dining room table. Have you looked there?" The child is then taught to change the subject, "Grandma, did I tell you about my school trip? We went to a farm and picked apples. We watched the people at the farm make apple butter. It was so much fun and it made me think of all the good things that you used to cook." This strategy of validating, offering to help, and changing the subject can be very effective and is one that school-age children can master.

Children may also worry that their parents are going to "catch" Alzheimer's like they catch colds from others. They need reassurance that Alzheimer's is an illness that is not spread from one person to another like a cold and that most people do not get the disease. It is important for children to be told that grandma will have good days and bad days and that the routines in the home might change. Most important, children need to be told that they are loved—no matter what the circumstances of grandma's illness.

It is important to encourage conversation about grandma. Parents should open the conversation by asking children about their own observations concerning grandma. Have they noticed anything that she really likes or that makes her smile? Have they noticed any changes in grandma and how do they feel about those changes? Helping children to talk openly about their own feelings, worries, and fears allows the parent to reassure the child, let them know that their feelings are normal, and that their parents also experience similar feelings.

Sometimes children express their emotions in indirect ways. A child may complain of "not feeling good" but not be able to pinpoint what is wrong. School performance might decline. Children may be hesitant to invite friends to their home because they are afraid of what grandma might say or do. These behaviors need to be gently confronted while offering the child comfort and support.

It is important to help children stay involved with their grandparents. They can be encouraged to set the table together, to look at picture albums, listen to music, dance, or engage in any simple activity that can be shared. It is important for parents to help children find ways to continue to show love to the grandparent with dementia.

Finally, it is helpful for parents to read to their children age-appropriate books on dementia. There are a number of excellent books on the market. For instance, *What's Happening to Grandpa?*(2004) written by Maria Shriver is a sad but wonderful book based on Shriver's experiences with her own father who died from AD. *Always My Grandpa: A Story for Children about Alzheimer's Disease,* written by Linda Scacco and Nicole Woy in 2005, and *Still My Grandma,* written by Veronique Van Den Abeele and Claude Dubois in 2007, are other excellent books that can help parents explain this terrible disease that is slowly taking grandma or grandpa away (Carson 2012).

Holidays

Every aspect of life changes when a family member becomes the caregiver of someone with AD. Although change does not necessarily mean giving up the traditions that have always held great meaning, it does mean that some traditions may need to be revised to accommodate the demands of being a full-time caregiver (Carson 2011). For instance, if the tradition has been that the caregiver cooked the entire holiday meal, then that will need to be modified to allow for wider participation in preparations for the holiday. Other family members will need to be more involved and/or the meal simplified so that the workload on the caregiver is minimized. Sometimes when faced with a dramatic change, like providing full-time care to a loved one with Alzheimer's, we are forced to reexamine the true meaning behind celebrations. Often the reason for the celebration is to share a meal with loved ones—does it matter if the menu is simple? Does it matter if the host ordered the meal from a local restaurant rather than trying to prepare it while taking care of a loved one's needs? Or is it not more important that a family gathers together?

Life changes in both small and large ways when a loved one has AD. The prevailing rule must be simplify, simplify, simplify! Adapting to these changes does not necessarily mean totally giving up activities that were once enjoyable. Changes can be made that keep the reason for the celebration in the forefront and relegate the details of preparation to the background.

One of the major difficulties, but one of the most important issues to confront, is that the power of traditions often locks us into ways that we have always celebrated the holidays—even when those traditions no longer make sense. It is important for the hostess or host to plan early and involve a "committee" of loving family and/or friends. It helps to begin with a list of all the chores that must be accomplished—followed by a careful review of that list in order to vigorously slash the "must-dos" to a reasonable "can-do" list. Keep in mind that this gathering is to celebrate the holiday (whichever holiday it is) and for loving friends and family to share a happy meal together and to enjoy each other's company. Once the purpose is cemented into the minds of the hostess/host and the planning committee, the idea of simplifying the planning becomes a fun challenge to see which member can do the best job of simplification of age-old traditions! Someone on this committee needs to take notes and share them with everyone after the meeting. Why would this be valuable? Because if this first plan is successful, it becomes the template for future successful events—the caregiver most likely will have less, not more, time to plan, celebrate, and get together with friends as the disease continues its devastation. This will also help the caregiver to obtain greater support from family and friends—so a good plan lessens the work later on and helps to cement a group of supporters who will be much needed as the disease progresses.

The plan needs to address the "who," "what," "how," "where," and "when" for the gathering:

1. "Who" will be involved in the planning and preparation for the celebration as well as who will be invited—keeping in mind that the person with Alzheimer's fares much better in small rather than large groups.
2. "What" food will be served—keeping in mind that finger foods might be the easiest for the person with Alzheimer's to handle—he/she can walk and eat and not worry about table manners that may have disappeared right along with her/his short-term memory.
3. "Where" will the celebration be held (keep in mind that the least stressful environment will be the environment most familiar).
4. "What" activities can be done for entertainment? As the planning proceeds, it is important to remember the power of music to engage those with AD. The person with dementia responds to music in powerful ways: by singing, dancing, and just tapping their feet. A sing-along based on a playlist of songs that might have been among the favorites of the person with Alzheimer's might be amazingly successful in encouraging participation. Additionally, it helps to print the lyrics of these favorites in a large font—most individuals with dementia can still read late into the disease, so having the words in front of them will encourage singing. Another activity that engages the person with Alzheimer's is to find old advertisements that the person might have been exposed to in their youth and early twenties—and have the group yell out the missing words to the advertisement. For instance, "You'll wonder where the Yellow went when you brush your teeth with…" (and the group yells out Pepsodent). These kinds of activities draw on old memories and allow the person with Alzheimer's to actively participate, have fun, and feel like part of the group. It is important for someone on the planning committee to save these activities (if they work) and help the person with Alzheimer's engage with the group. This will increase the chance that these same activities will work over and over again. The person with Alzheimer's will not become bored because he/she lives in the immediate present and will not remember engaging in these activities even in the recent past.
5. "How" to assist the person with Alzheimer's to feel as if she/he belongs to the gathering—this involves using their remaining reading ability. One of the planning committee needs to be responsible for creating large name tags: "My Name Is," and include relevant talking points about that person. For example, "My Name Is Mary Jones—we have been friends for 30 years; recently, my oldest son Adam and his wife Shelley had their third baby—her name is Marleigh. Marleigh has two older brothers—Gavin and Tanner." This provides the person with dementia with not only conversational starters but also ways of orienting him/herself towards others.

These are not difficult strategies and approaches. If they are carried out with the support of a committee of loving family and friends, any burden that is involved is minimized. Continuing to celebrate is essential to being fully human, and when someone is no longer able to initiate celebrations, we need to take over for them—by celebrating holidays, birthdays, and perhaps even some "made-up" special days. These are ways of continuing to love not only the old person that caregiver has known but also the changed person who continues to live and at some level maybe in desperate need of just being in the middle of a celebration and happy times (Carson 2011).

Gift Giving

What to give the person with dementia for birthday and any celebrations that routinely include exchanging of gifts can be a major issue for the gift-giver. The stage of dementia will impact choices made about gifts. If someone is in stage 4 of the FAST scale—gift giving will not be dramatically impacted. The person will still be functioning quite well, and when friends and family look back, they will remember how well the person seemed to be doing. Remember, this is the stage of the "Great Fooler." Stage 5 presents a different situation. At this stage, those with dementia are much more impaired. They can no longer shop independently, clean their homes or apartments, make decisions about what to wear, or handle money. Gifts might include providing cleaning services for the home, the automatic delivery of food and other necessities from a local super market, or the purchasing of sweatpants and tops that are easy to put on and easily washed. During stage 6 and beyond, those with Alzheimer's or another dementia may not be able to recognize gifts so that keeping them practical and simple are important. Creating a memory book with pictures of family members along with names and relationships written under each picture would be wonderful gifts that will help the person feel more oriented than they actually are. Creating a video of the family where each person introduces himself or herself to the person with Alzheimer's is a powerful "memory gift." Lastly, creating a music library on an iPod of songs that were popular when the loved one with dementia was in his/her early twenties back to childhood will enable the person to enjoy music from his/her era. In stage 7, the person will be unaware of holiday celebrations but will still respond to music. The caregiver, in contrast, will still be fully aware and will need to grieve the loss of holidays and celebrations. Family and friends can make these times special for the caregiver—by either relieving her/him from the daily responsibilities of caregiving or bringing all elements of the celebration to the caregiver and the person with dementia.

Making the Painful Decision: Placement of a Loved One

There is nothing kind about dementia. It is like a burglar that steals a person's most precious possessions and tosses them out as if they were trash—memories, relationships, one's history, one's present, and one's dreams and achievements. They are all for naught or this would seem to be so. The disease demands an inordinate price at the end, as all the functions of an independent person slip away and they begin to function at the level of an infant. Could there be anything crueler? But indeed, there is still more that must be endured. Family caregivers are exhausted by lack of sleep and incredibly weary over trying to be caring and loving. End-of-life decisions, which are never easy, must be made by those caregivers who have faithfully ministered to a loved one perhaps over many years.

Ideally, the person with dementia has made their wants known with regard to how they want to die and completed an advanced directive to guide those painful end-of-life decisions. Hopefully, the primary caregiver is not left to make end-of-life decisions without the support of other loving family members. A good support system allows the primary caregiver to function at their best when it comes to helping a loved one with dementia die.

Early-Stage Preparations

There are some preparations that are best done sooner rather than later. It may be hard to consider these questions at first, since it means thinking about a time when a loved one will be well down the road of his or her Alzheimer's journey. However, putting preparations in place early helps to make the transition smoother for everyone. Depending on the stage of diagnosis, include the person with Alzheimer's in the decision-making process as much as possible. If their dementia is at a more advanced stage, at least try to act on what their wishes would likely be.

Questions to consider in early preparation:

- *Who will make health-care and/or financial decisions when the person is no longer able to do so?* While a difficult topic to bring up, if the person with dementia is still lucid enough, getting her/his wishes down on paper means they will be preserved and respected by all members of the family. Meeting with an eldercare law attorney is the best way to learn about options. Issues that are important to consider include obtaining power of attorney for both finances and health care. If the person has already lost the capacity to make competent decisions, the primary caregiver will need to apply for guardianship/conservatorship.
- *Who will provide care?* Sometimes, family members assume that the spouse or nearest family member will take on caregiving responsibilities, but that is not always the case. Caregiving is a large commitment that gets bigger over time. The person with dementia will eventually need round-the-clock care. Family members may have their own health issues, jobs, and other responsibilities. Communication is essential to make sure that the needs of the Alzheimer's patient are met, and that the caregiver has the support to meet those needs.
- *Where will the person live?* Will remaining in his or her home be possible, or will the home be difficult to access and hard to make safe for later? If the person is currently living alone, for example, or far from any family or other support, it may be necessary to relocate or consider a facility with more support.

It is important to find out what assistance is available from the medical team. Increasingly, families are enlisting the services of a geriatric care manager, who is usually a nurse or a social worker. Geriatric care managers can provide an initial assessment as well as assistance with managing details of placement in an assisted living facility or a nursing home if that becomes necessary later.

Considering Long-Term Care

As noted repeatedly, it is the nature of AD to progressively worsen as memory deteriorates, and in the advanced stages, round-the-clock care will be necessary. It is important to think ahead and plan for this. The following are possibilities to consider.

Care at Home

There are several options for providing care at home:

- *In-home help* refers to hiring private duty caregivers who can provide in-home assistance to a loved one. This can range from a few hours a week of assistance to live-in help, depending on the needs of the caregiver and the person with dementia.
- *Day programs, also called adult day care,* are programs that typically operate weekdays and offer a variety of activities and socialization opportunities. They also provide the chance for you as the caregiver to continue working or attend to other needs. There are some programs that specialize in dementia care. Most provide transportation to and from the facility. Some provide bathing services as well.
- *Respite care.* Respite care is short-term care where a loved one stays in a facility temporarily. This gives the family caregiver a block of time to rest, travel, or become involved in other things that provide relief and respite.

Is It Time to Move?

As Alzheimer's progresses, the physical and mental demands on the caregiver will gradually become overwhelming. Each day will bring more challenges. The individual with dementia may eventually require total assistance with physical tasks like bathing, dressing, and toileting, as well as greater overall supervision. At some point, the caregiver will not be able to leave her/his loved one alone. Nighttime behaviors may interfere with the caregiver's sleep, and if the person with dementia is belligerent or aggressive, this may exceed the caregiver's ability to cope or feel safe. Sometimes, bringing in additional assistance, such as in-home help or other family members to share the caregiving burden, will enable the caregiver to keep the loved one at home. However, it is not a sign of weakness if the caregiver concludes that moving the loved one to a facility is the best for both the loved one and the caregiver. Placement of a family member is always a difficult decision to make, but when the caregiver is overwhelmed by stress and fatigue and her/his own health problems, it is time to consider placement.

Assisted Living Facilities and Nursing Homes

Assisted living is an option for those who need help with some activities of daily living, but not others. Certain facilities may also provide assistance with medications. Staff is available 24 h a day, but it will be important to know if the staff has experience handling residents with AD. It is also important for the family caregiver to communicate what stage of dementia the loved one is in. The facility may not be able to provide the level of care needed and the loved one may need to be placed in a skilled nursing facility also called a nursing home.

Nursing Homes

Nursing homes provide assistance with both activities of daily living, medications, and constant monitoring. A licensed physician supervises each resident's care and a nurse or other medical professional is almost always on the premises. Skilled nursing care providers and medical professionals such as occupational or physical therapists are also available.

Making the Choice of Facility

Once the caregiver has determined the appropriate level of care, it will be important to visit the facility—both announced and unannounced—to meet with the staff and determine the level of skill and experience the staff has with dementia residents. Facilities that cater specifically to Alzheimer's patients should have a designated area for care, often called a *special care unit.* It is important to ask questions of the facility's management and staff and to make observations. For example:

- Does the facility mix patients with dementia together with those who have psychiatric illness? This can be a dangerous practice.
- Does the program require the family to supply a detailed social history for the resident? This is a positive practice.
- Is the unit clean? Is the dining area large enough for all residents to use it comfortably? Are the doors alarmed or on a delayed opening system to prevent wandering? Is the unit too noisy?

- Does the unit use paint and colors in a way that assists the dementia patient to be as independent as possible. For instance, bathrooms should have contrasting colors—if everything is white, including the walls and floor, the person with dementia will not "see the toilet."
- Are there activities provided every day? Is music part of the activity plan?
- Is signage used to help residents be as independent as possible—taking advantage of the fact that those with dementia can read until very late in the disease.
- What is the ratio of residents to staff? (5:1 during the day, 9:1 at night is normal). What is staff turnover like? How does staff handle meals and ensure adequate hydration, since the person with dementia can often forget to eat or drink? How do they assess unexpressed pain if the resident has pain but cannot communicate it?
- What training for Alzheimer's care do they have? Does the facility provide staff with monthly in-service training on Alzheimer's care? Do they hold case conferences to discuss the management of challenging behaviors?
- Is there an activity plan for each resident based on the person's interests, life story, and remaining cognitive strengths? Are residents escorted outside on a daily basis? Are regular outings planned for residents?
- Does the unit provide hospice services? What were the findings in the most recent state survey?

Making the Move

Moving is a big adjustment both for the person with Alzheimer's and the primary caregiver. The loved one is moving to a new home with new faces. The caregiver is adjusting from being the hands-on caregiver to being an advocate. This adjustment can be difficult, and it takes time. Moving day can bring with it a variety of emotions—sadness, guilt, and relief are common for the caregiver, and agitation, uncertainty, and fear for the loved one with dementia. Adjustment will come with time.

Each person will deal differently with this transition. Depending on the loved one's needs, the caregiver may either need to visit more frequently or lessen visits in order to give the loved one her/his own space to adjust. As adjustment proceeds, the caregiver can settle into a visiting pattern that is best for both the person with dementia and themselves.

Summary

There are many issues that arise in the care of someone with dementia. This chapter presents possible solutions. This includes keeping good records that enable caregivers to recognize patterns in behavior and lead to problem solving. Taking away the car keys is a huge issue for the person with dementia who associates driving with adulthood and independence, but also for the caregiver who worries about safety issues. Sundowning is a challenge difficult to management. The approach presented in this chapter is based on the theory of retrogenesis. Explaining to grandchildren about why their grandparent no longer seems to know them is a sad reality that must be faced and explained so that children can understand and accept the situation without feeling devastated. Vacations that include traveling with someone with dementia present unique challenges—regardless of whether the travel is by car or plane. Each mode of travel requires special considerations and preparation. Handling holidays can be another hurdle for families, especially if they are bound to traditions. When someone in the family has dementia, traditions may need to be altered to accommodate the person's needs and capabilities. Finally, placement decisions are painful to make, and this chapter provides guidelines on how to make that decision.

References

Carson, V. B. (2011). Alzheimer's disease and holiday celebrations: may not be the holidays you remember. *Caring Magazine, 30*(12), 54–55.

Carson, V. B. (2012). On the road again … taking a vacation with a loved one with Alzheimer's disease. *Caring Magazine, 31*(8), 40–41.

Carson V. B. (2012). What to say to grandchildren who ask, "what is wrong with grandma or grandpa?" *Caring Magazine, 31*(9), 46–47.

www.best-alzheimers-products.com.

The Hartford, Warning Signs for Drivers with Dementia. www.thehartford.com/the hertford/.../dementia-warning-sings.pdf

www.alzheimersvideo.com/

Chapter 11
Finale: Pulling It All Together

In the introduction of this book, we mentioned that the techniques and approaches that are suggested are examples of thinking outside of the box. Joyce's story is a powerful example of how she and her family were able to think outside of the box. What does this mean? Take a look at the two diagrams below—the only way to connect the nine dots using only four lines is to literally think outside of the box!

We Need to Think Outside of the Box!!

Can you join all the points together with only four lines?

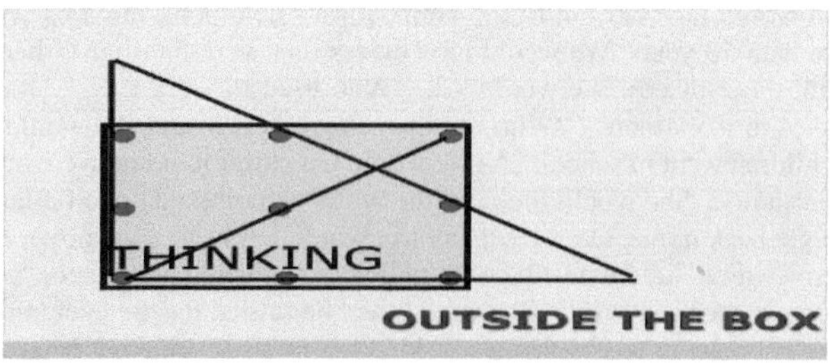

Keep this in mind as you read Joyce's story.

This final chapter is a love story of sorts between a mother, Jane Kontrabecki, who had Alzheimer's dementia, and her daughter Joyce Markowitz who provided care to her mother and assisted the rest of the family to do the same. Joyce is a nurse and a chief executive officer (CEO) of Long-term Care for the Catholic Health Care System in Western New York. The care that Joyce and the rest of the family pro-

V. Benner Carson et al., *Care Giving for Alzheimer's Disease*,
DOI 10.1007/978-1-4939-2407-3_11

vided was patient, loving, and individualized. It was based on Jane's life story—the activities she enjoyed doing, the way she liked to bake cookies and cakes, the way she liked to celebrate holidays, the significance of her strong religious beliefs—all of these things that the family knew about their mother and grandmother were lovingly woven into her daily care. Although the family was not confronted with every challenging behavior described in this book—they dealt with many of the behaviors in a creative and loving fashion. Much can be learned from those who have walked this journey of care. Let us listen to Joyce's story that she shared with the authors on November 2, 2011.

> First I would like to talk about the fact that I am a nurse and when my Mother first started exhibiting symptoms, I would often address those symptoms based on my training as a nurse. For instance if my mother was confused I would attempt to reorient her to time and place. I learned however that as her disease progressed that most of what I needed to do was very intuitive as to what you would do in a loving relationship. I always made certain for example that when I spoke with her that I pulled myself close so that we were touching. She often appeared to be very lonely even when others were around, but holding her hands and reassuring her through touch was very important especially when she was agitated.
>
> I have a habit with people that I love: I usually greet them by placing my hands on each side of their face. I always greeted both my Father and Mother this way so this was a very familiar form of recognition for my mother and very powerful when she was disoriented. When I came into the room and touched her face she always knew it was me and I think the familiarity of this gesture was the reason she was never confused about my identity.
>
> My mother's home was very important to her: she never wanted to leave her home even after my father passed away. One of his concerns was that she would be moved away from the home where she raised her children and he knew that if we were to do so that this would be devastating to her. Although her home was her place of stability she would frequently become confused about where she was and she would ask that we take her home. When she initially exhibited this behavior I would often be able to calm her agitation by showing her the photographs in the house and ask her to identify the faces so she could associate the people with her home. As her disease progressed this was no longer a viable solution, and when she thought she wasn't home she would be obsessed with that throughout the evening often crying and begging for someone to take her home. When this behavior occurred my daughter and I would take her out of the house for a ride around the area where she had lived for more than 50 years. We would pass places that were familiar to her and ask her questions about these places. We would ask, "Who used to work there?" and she would reply, "Oh I used to work there." "Who went to school there?" and she would reply, "That is where my children went to school." As we would get closer to home we would ask her to help us find her house. She would look out the window, point out specific landmarks and tell us how to get back home. On a good day she would tell us to slow down and point out the specific driveway to her home. Once we pulled in we would announce, "we are home" and we would take her back inside. She would then be settled for the evening and realized she was home. There were weeks that we needed to do this seven days in a row but it was a ritual that brought her peace and calmness so we continued to take her for these rides.
>
> Growing up my Mother was known for being a great baker. When we were children she would get up very early on Sunday morning and make enough dough for six loafs of bread, dozens of dinner rolls, cinnamon rolls and coffee cakes. You could smell the aroma of the sweet dough baking throughout the house. She loved to cook, bake and make her own jelly and jams. My daughter who inherited these skills from my Mother began baking with her at least once every week. My daughter would pre-measure the ingredients of recipes that my mother and her grandmother had baked for many years and place all the needed utensils within mother's reach. My mother would pour all the ingredients into the bowl while my

daughter cracked the eggs and dealt with the tasks that presented more of a challenge. At Christmas we followed the same approach but would make familiar recipes that also challenged her manual dexterity. We would make dough for example that needed to be rolled in the palm of her hand. We would scoop the dough and let her roll it into a ball and place it on the cookie sheets. We also would allow her to place the sprinkles on the cutouts. This created a bit of a mess but she enjoyed being able to do something that she had done for so many years. Simple repetitive tasks that she was able to achieve brought back what she had so enjoyed doing when she was healthy. These were the moments when we would see little bursts of her "old self" where she would actually be able to engage with us; she would be more oriented and she would seem like herself, even if it only lasted for a short time.

Her relationship with my daughter as well as my sister, who spent a great deal of time with her, was so important during her periods of confusion. She exhibited Sundowning episodes that made her very confused and disoriented. She would often call out at night for my Father who had passed away and it was during these times that she needed the familiarity of her daughter or granddaughter to provide her with comfort.

My mother had six children and no matter how confused she became, she was always a mother at heart. When her Grandchildren began to start their families they would bring over their children and she would sit in the chair and automatically start rocking, and patting the baby. Years of being a mother would surface and bring her such joy. As often as possible we gave her the opportunity to be with children.

When she was confused at night, my daughter would climb in bed with her, cuddle close to her and tell her, "Grandma, I'm your baby." At that time my daughter was 22 years old but my Mother would pull her close as if she was a small child. They would sleep with each other and my Mother would be at peace. My mother was always a loving woman who demonstrated that love through acts of affection toward her children and grandchildren so when she was the most agitated we used touch to help her settle down.

We never left my mother alone and therefore we frequently enlisted the help of outside caregivers. I lived within walking distance of my mother and there were times when the caregivers would call seeking advice during an outburst or periods when she would cry. I found that going back to what was familiar and important would work, not only for family but also with those less familiar. I would ask the caregivers to begin to recite the Lord's Prayer or sing a familiar hymn and no matter how agitated, this would distract her from the behavior and she would participate in the prayer or the song.

Saturday was our special day and everyone who knew this would remind her that I would be coming over. I would arrive before she was even awake, prepare breakfast and wait for her to awaken. We would start out with breakfast and then I would bathe her. I would give her a head to toe inspection during this time looking for skin breakdown, bruising or any injuries that she may have sustained.

We were fortunate that my brother had the financial resources to have my mother's bathroom remodeled to accommodate her care. We removed the tub and placed a hand held shower and shower chair in the bathroom. This worked extremely well since she could be involved in the bathing process. She didn't like having her hair washed and I was the only one she would allow to wash her hair. She was very fearful of having the water go into her eyes so I would give her a towel and have her put it near her face. She would tip her head back and I would rinse the soap from her hair. During bathing she had to be in control. I would let her hold the shower head and she would often turn and squirt me and I would laugh and she would start to laugh as well. Most of the time she didn't realize she was squirting me. I would say "look at me! I am drenched, we are supposed to be giving you a shower not me!" And she would laugh even louder.

I found that she became cold very easily, even in the warm weather, so we kept the bathroom very warm. This was important because at times she needed coaxing in order to be bathed and being cold could not be used as an excuse. On days that she really did not want to be washed, I would need to be firm. She was frequently incontinent so I would remind her that she always took such good care of herself, so I would say, "You do not smell very good today and we don't want anyone to think that you don't smell good, so let's get you washed up." That was usually enough for her to allow me to bathe her.

Once we were ready to go we would devote the rest of the day to an adventure. We would do things like go to the grocery store and shop together. She wasn't very mobile so I would use one of those carts that allowed her to sit and I would push her around the store. She always enjoyed grocery shopping so we would take our time to touch and smell. I allowed her to make choices about what she wanted to take home. We always took home a bouquet of flowers.

Sometimes we would use Saturdays to recreate the rides that she had taken with my Father. I would put on her special sunglasses that we purchased just for our outings and head to the country where we would stop at the roadside stands and purchase berries or any produce in season. Our final stop was usually lunch and then we would head home where she would take a very much needed nap.

My Mother probably had Alzheimer's for almost ten years. She had early symptoms that we didn't even recognize like forgetfulness, and then she suffered a stroke and everything began to accelerate. She was seeing the same primary care physician for many years and like many long standing relationships I found that the primary care physician often could not see many of the subtle changes since he knew my parents for so long. He often would ask me what I wanted him to do. He was more apt to treat her elevated blood pressure than address anything to do with her dementia. I think because he knew that her children were such strong advocates for her, he just let us handle things on our own.

She would frequently suffer from urinary tract infections or would become dehydrated. We were constantly pushing fluids but with the UTIs and dehydration came an entire set of behavioral changes. Her complete demeanor would change and she would either become extremely aggressive or the exact opposite—she would be very lethargic. These extreme behavior swings became our trigger for medical intervention. We quickly learned however that a hospital admission was never a good experience. The disruption to daily routine, lack of ambulation and occasionally an unrequested dose of Haldol would result in weeks of needing to recondition her at home. We learned to medically manage her as an outpatient whenever possible, thus improving her quality of life.

One of our most challenging experiences was related to the holidays. When I was growing up Christmas was such a huge holiday for her. The more presents her children opened the happier she would be and so she wrapped everything. She would wrap a box of Crayola Crayons, a coloring book and any item that would add to the pile. Gift giving was so important to her, so when Christmas would approach she would become upset because she was unable to shop for gifts. She had one sister Alice and it was so important for her to give Alice a gift. If she was up to it I would take her out to purchase a gift for Alice but as her health deteriorated I would purchase several items and let her choose. I would ask, "Which do you think Alice would like the best?" and she would pick it out and then we would wrap it together, even if she just held the tape. We would put the gift under the tree and when Alice came by Mother would have a gift to give her only sister. She had me do the same for my brother's children. They were the last children to be born with a large age difference from those of her other Grandchildren. Although they lived far away she wanted each of them to have a gift from their Grandma.

On the last Christmas Eve that we were together she came to my home as she had done for the past 30 years. She was very agitated all evening and was convinced we were all lying to her. I can't even remember what it was about, it was something mundane, but she kept screaming, "You're lying, you're lying." I took her home and stayed with her until she settled down. We had a cup of tea in the quiet of her home and I told her that I had never lied to her and never would; she didn't really remember what had happened that Christmas Eve. However I learned that going forward we needed to celebrate in her environment and with less people. Too much noise and activity even when it was family had become overwhelming for her. We took Christmas to her house, cooked in her kitchen, ate at her table with her dishes and with a smaller group. All future festivities were done in her home and when she was able she peeled potatoes, held the hand mixer or folded table napkins. Whatever she could do to be engaged she was invited to do and we this proved to be far better than taking her out of her environment.

I learned that caring for someone you love with dementia must be done from the heart. It doesn't require complex thinking but simple things, common sense. We had a home health aide that always reminded my mother that my father had passed away. What she didn't realize was that every time she told my mother it was as if she was hearing it for the first time. I had to explain this to her and suggest that she tell my mother that dad was fishing. Dad loved to fish so my mother was always ok when that was the response to, "where is Joe?"

One rainy night my mother would not stop crying. I received a frantic call from the home health aide who had tried many of the techniques that had worked in the past but without success. When I arrived at the house my mother quickly told me that she had no money to take the bus and that she needed bus money. My mother had not been on a city bus since she was a young girl but this was important to her. I found a small change purse, put some coins inside and gave it to her to hold. It was dark outside and I told her it was better to wait and catch the bus in the morning. She agreed and by morning did not remember a thing. As much as I wanted to reason with my mother I knew that I must come into her world and when I did so I was careful to be respectful and treat her as my mother. She would sometimes say to me "don't talk to me like a baby". Those were the moments that I knew she was in there and that she was struggling with the fact that something was wrong that she couldn't control.

The day my mother passed will forever be in my memory. Her heart rate suddenly slowed and the look on her face was filled with fear and uncertainty. We gathered those that loved her to her bedside but she only looked at us with eyes that were filled with uncertainty. That blank look that so often accompanied her confusion. In those moments I followed that same instinct that had become my compass for caring for this wonderful woman. I climbed in the bed and held her in my arms and whispered that I would not let her go. She finally closed her eyes and passed within a few minutes. I believe that I alleviated her fear and I hope it was that loving human touch that allowed her to leave this earth in peace.

Appendix

V. Benner Carson et al., *Care Giving for Alzheimer's Disease*,
DOI 10.1007/978-1-4939-2407-3

POLICIES AND PROCEDURES CONCERNING SEXUAL EXPRESSION AT THE HEBREW HOME AT RIVERDALE

Updated by

Robin Dessel, LMSW
Sexual Rights Educator

Mildred Ramirez, PhD
The Research Division

The Hebrew Home at Riverdale
Riverdale, New York

August 1995
May 2001 (revised)
April 2013 (revised)

INTRODUCTION

The Hebrew Home at Riverdale ("the Home") recognizes and respects the importance of emotional and physical intimacy in the lives of older adults. Such close human interactions are viewed as a normal and natural aspect of life.

Planning for the social needs of older adults, including those with Alzheimer's disease and dementia, carries with it a responsibility to uphold their personal and sexual choices. This Sexual Rights Policy applies to all older adults receiving services from or residing at the Home (i.e. tenants, clients, participants), hereafter referred to as "residents".

The Home's policy and practice is based on the assumption of autonomy, civil and privacy rights of all people, including the protected rights of all older adults residing in Medicaid/Medicare settings.

This policy recognizes and supports the older adult's right to engage in sexual activity, so long as there is consent among those involved. Consent may be demonstrated by the words and/or affirmative actions of an older adult:

- **a) with intact decision-making ability; or**
- **b) with intact decision-making ability who is non-verbal; or**
- **c) with Alzheimer's disease or Dementia**

The former may require, and the latter does require an assessment conducted by clinical staff, using the Home's "Assessing Consent to Sexual Activity" guidelines[1], to confirm that consent was and continues to be given, and to insure the safety of those involved.

It is recognized that all adults are entitled to their personal beliefs, choices and practices in matters of sexuality. In assessing sexual activity in the context of this policy, professionals and caregiver staff, however, should set aside personal biases and judgment to maintain objectivity in upholding the sexual rights and choices of all older adults in the Home's care.

Following is the translation of these tenets into a set of Policies and Procedures which serve as the basis and reference for sexual expression at the Home.

---

[1] Available on Hebrew Home website @ www.hebrewhome.org

A. Resident Sexual Rights

Preamble A
For the purposes of the following rights and responsibilities, sexual expression is defined as:

"Words, gestures, movements or activities (including reaching, pursuing, touching or reading) which appear motivated by the desire for sexual gratification"

Residents have the right to sexual activity providing in each instance that the activity does not involve:

(1) non-consensual acts
(2) acts with minors
(3) acts between persons if there is any possibility of the transmission of an STD
(4) acts that impact negatively on the resident community as a whole through public display

In the case of older adults with physical and/or cognitive impairment, the above noted provisions apply in addition to an assessment by the clinical staff to insure consent.

Rights

1. Residents have the right to seek out and engage in sexual expression as defined above.

2. Sexual expression may be between or among residents, and may include visitors subject to the conditions expressed in Preamble A. Sexual expression between a resident and visitor, which is arranged by a resident and/or family, is further subject to the stipulation that it is a legal arrangement; and there is no solicitation of other residents.

3. Residents have the right to access and/or obtain, for private use, materials with legal but sexually explicit content: books, magazines, film, video, audio, pictures, or drawings.

4. To the extent possible, residents have the right of access to facilities, most notably private space, in support of sexual expression.

5. Residents have the right of access to professional counseling pertaining to sexual expression of self or others.

In instances where conditions expressed in Preamble A are not fulfilled, including those involving cognitively impaired residents, the relevant Interdisciplinary Care Team (ICT) will make clinical determinations weighing the relative benefits or potential harm associated with the resident's(s') sexual expression. Involvement of a family member or legal representative may be indicated only in instances where the involved resident(s) is cognitively impaired. In such cases, it is the facility's responsibility to uphold the choices and rights of cognitively impaired residents, and to work with families and/or legal representatives if their suggested course of action is discrepant.

B. Staff Responsibilities

Preamble B
Staff should maintain an awareness of the Home's mission and policy, which respects the centrality of intimacy and privacy as a foundation for sexual experience and expression. Generally, it is the function and responsibility of staff to uphold and facilitate resident sexual expression.

Specific Responsibilities

1. Staff is responsible for maintaining an awareness of sexual expression, so as to ensure conformity to Preamble A.

2. Staff shall not intervene in resident sexual expression, except in instances where:

- **a. the conditions expressed in Preamble A are likely to be violated; or**
- **b. interventions recommended by the ICT are not followed.**

  In such instances and pending ICT care planning, staff interventions will emphasize diversion rather than prohibition.

3. Staff is responsible for recommending appropriate use of private space to the extent that it is available.

4. Staff is responsible for providing or obtaining consultation regarding sexual expression to residents upon request.

5. Staff is responsible for referring cases of sexual activity not covered by this policy, so that appropriate amendment can be made to both policy and practice.

C. Organizational Responsibilities

Preamble C
Just as residents have certain rights and staff certain responsibilities with respect to sexual expression, so the Hebrew Home has certain associated responsibilities, as follows:

1. The facility should ensure residents' privacy rights and, to the extent possible, make arrangements which facilitate sexual expression, e.g., a private room for one member of a couple.

2. The facility should develop and maintain an orientation and sensitization training program, in order to equip all staff with the skills necessary to upholding the sexual rights of all residents.

3. Through administrative directive and clinical supervision, the facility should ensure that the procedures outlined in this policy, are communicated to and followed by all staff.

ACKNOWLEDGEMENTS

With special appreciation to the policy's original authors

THE SEXUALITY WORKGROUP AT
THE HEBREW HOME AT RIVERDALE
August 1995

Douglas Holmes (Chair), Susan Fairchild, Robin Dessel, Nelson Burros, Ashley Coward, Lisa Howard, Irving Greenberg, Marjorie Labue, Janet Lowe, Paulette Rosner, Lauren Saul

With special thanks for their support in policy update

The Hebrew Home Administration

The Weinberg Center for Elder Abuse Prevention at the Hebrew Home

ABUSE or INTIMACY

Older Adult Sexuality

Aging does not mean a loss of sexual intimacy, but when an older adult is cognitively impaired (Alzheimer's, dementia) it can be difficult to determine whether he/she is engaged in a healthy sexual relationship or is a victim of a sex crime.
Is there an in between?
What is it and how do you know?
What happens when a 3rd party gets involved?

Older adults should be actively involved in the assessment process to protect their rights and choices

1. Has a crime occurred?
Example: rape or forced sexual act by a perpetrator of any age
Report to law enforcement immediately if a violation of the law is suspected. Each state has specific laws defining sex crimes.

2. Has sexual abuse occurred?
Example: non-consensual sexual act between vulnerable older adults; no clear intention by perpetrator to commit sexual offense
Follow regulatory reporting requirements even if there is no violation of a specific law.

3. Has inappropriate sexual activity occurred?
Example: hyper-sexuality or poor impulse control in older adult with dementia
Intervention and care planning is required to prevent inappropriate conduct, even when a law or regulation has not been violated.

4. Has a real relationship occurred?
Example: sexual activity which appears consensual between older adults with cognitive or physical conditions
No intervention is required when older adult is able to express some evidence of consent to participate in a healthy relationship. Monitoring is advisable to ensure continued consent.

The HebrewHome at Riverdale
SEXUAL RIGHTS PROGRAM

The Harry & Jeanette Weinberg
CENTER FOR ELDER ABUSE PREVENTION

Assessing Consent to Sexual Activity in Older Adults

1. Ability to express choices/consent

Ask:

- What are your wishes about this relationship?
- Does your sexual partner make you happy?
- Do you enjoy sexual contact?

Consider:

- Observations and non-verbal clues when older adult is unable to verbalize choices (facial expressions and body language)
- Emotion and mood, before and after sexual contact

2. Ability to appreciate sexual activity

Ask:

- Do you know what it means to have sex?
- What does it mean to you/your partner?
- What would you do if you wanted it to stop?
- What if your partner wanted it to stop?

Consider:

- Nature of the relationship (monogamous)
- Emotion and mood, before and after sexual contact

3. Personal quality of life choices in the here and now

Ask:

- Was and is intimacy important in your life?
- What are your social and companionship needs?
- What brings happiness or fulfillment to your day?

Consider:

- Past and present relationships (including family)
- Impact of cognitive impairment (not an automatic reason to deny relationship)
- Privacy and intimacy rights
- Responsibility to uphold older adults' choices
- Policies for staff education and practice
- Impact of third party objectives or values on assessment process

Index

V. Benner Carson et al., *Care Giving for Alzheimer's Disease*,
DOI 10.1007/978-1-4939-2407-3

FSC
www.fsc.org
MIX
Papier | Fördert gute Waldnutzung
FSC® C083411

Kolibri 360 GmbH
Ferdinand-Jühlke-Straße 7
99095 Erfurt, Deutschland
produktsicherheit@kolibri360.de